ADVANCED MAGNETIC NANOSTRUCTURES

ADVANCED MAGNETIC NANOSTRUCTURES

By

K.P. Awasthi

2008

CYBER TECH PUBLICATIONS
4264/3, Ansari Road, Daryaganj, New Delhi-110002 (India)
Ph.: 011-23244078. Fax: 011-23280028
E-Mail: cyberpublicationsdelhi@yahoo.com

ADVANCED MAGNETIC NANOSTRUCTURES

K.P. Awasthi

First Edition 2008

Published by :

G.S. Rawat for Cyber Tech Publications
4264/3, Ansari Road, Daryaganj, New Delhi-110002 (India)
Ph.: 011-23244078. Fax: 011-23280028
E-Mail: cyberpublicationsdelhi@yahoo.com

Printed at : **Amit Enterprises**, Maujpur, Delhi, (9810878184)

Preface

Advanced magnetic nanostructures is an emerging field in magnetism and nanotechnology, but the literature consists of a rich variety of original papers and parts of reviews and books whose scope is comparatively broad. This calls for a book with specific emphasis on state-of-the-art synthetic methods for fabricating, characterizing and theoretically modeling new magnetic nanostructures. This book is intended to provide a comprehensive overview of the present state of the field. Leading researchers world-wide have contributed a survey of their specialties to guide the reader through the exploding literature in nanomagnetic structures.

The focus is on deliberately structured nanomagnets. It includes cluster assembled, self-organized and patterned thin films but excludes, for example, multilayered thin films. We target both industrial and academic researchers in magnetism and related areas, such as nanotechnology, materials science, and theoretical solid-state physics.

This book has been written to make readers inform about the advancements and technological development in the field of Magnetic Nanostructres. I sincerely hope that this book will prove to of great worth to the readers keen to know about the magnetic nanostructures.

Your valuable suggestions and comments are always welcome.

—*Author*

Contents

1

Introduction

The nanostructures considered in this book are magnetic and characterized by *structural* length scales ranging from a few interatomic distances to about one micrometer. The basic length unit is the *nanometer* (1 nm = 10–3ìm = 10–9m), corresponding to about four interatomic Fe-Fe distances. Magnetic nanostructures pose experimental challenges, exhibit interesting physical phenomena, and have many present or potential applications. An important aspect is that structural lengths affect, but only partly determine, the magnetic length scales encountered in the structures. Examples are domains in semihard nanoparticles, where both the domain size and the domain-wall thickness may be smaller than the particle size, and polycrystalline soft-magnetic nanostructures, where the magnetic correlation length is much larger than the crystallite size.

Due to rapid progress in the fabrication and processing of nano structures, it is now possible to realize a broad variety of geometries, crystalline textures, and chemistries. For a given geometry, the structures can be fabricated using a variety of magnetic materials (compounds), with different local magnetic properties and crystalline textures. The Appendix presents some magnetic materials of interest in nanomagnetism.

Basic Definitions and Units

The magnetic moment m of the atoms in a nanostructure nearly exclusively originates from the electrons in the partially filled inner shells of transition or rare-earth metals. There are both spin (*S*) and orbital (*L*) contributions, but since L is much smaller than S in most iron-series transition-metal magnets, the magnetic moment is often equated with the spin polarization. The situation is similar to that encountered in bulk magnets, although both *S* and *L* may be modified at surfaces and interfaces (Ch. 2). As in infinite solids, nuclear moments are much smaller than electron moments and can be ignored safely for most applications.

The magnetic moment of an atom is created by intra-atomic or Hund's-rules exchange, which favors parallel spin alignment on an atomic scale. In addition, ferromagnetism requires *interatomic* exchange, to ensure parallel alignment of the moments of neighboring atoms. The resulting net moment gives rise to the magnetization M = J/ìo which is defined as the moment per unit volume. In bulk ferromagnets, the competition between interatomic ex change and thermal disorder leads to the vanishing of the spontaneous mag netization *MS* = | M(r) | at a well-defined sharp Curie temperature *T*C. How ever, the existence of a Curie-point singularity is limited to infinite bodies, and in nanostructures, the concepts of magnetic phase transitions must be reevaluated.

Spin-orbit coupling in combination with local crystal-field interactions gives rise to magnetic anisotropy. In the simplest case, magnetic anisotropy is of the uniaxial type and described by the lowest-order anisotropy constant *K*1.

This constant is equal to the energy density necessary to turn the mag netization from the easy to a hard magnetization axis. In addition, there are shape-anisotropy contributions of magnetostatic origin. In nanostructures, surface, inter-face, and shape anisotropy contributions are often important, particularly in cubic materials, where the lowest-order bulk anisotropy is zero.

An applied magnetic field H changes the magnetization by rotating the local moment. Since the magnetic anisotropy yields energy barriers between different magnetization states or spin configurations M(r), the field dependence of the net magnetization exhibits *hysteresis.* Important hysteretic properties are remanence *M*r, defined as the zero-field magnetization after saturation in a strong magnetic field, and the coercivity *H*c. The latter is defined as the reverse magnetic field at which the volume-averaged magnetization of an initially saturated magnet reaches zero. Some other hysteretic properties of specific nanostructuring offers many tools to tune hysteretic properties. For example, the coercivity of advanced magnetic nanostructures varies from about 1 ìT to several T.

Analytical and numerical aspects of the hysteresis of magnetic nano-structures. The structural length scales of nanomagnets are intermediate between interatomic and macroscopic distances, but nanomagnetism cannot be reduced to a mixture of atomic-scale and macroscopic phenomena. For example, most extrinsic properties are realized on a nanoscale, and nano structuring is used to produce optimized hard, soft, information-storage and sensor materials. A dynamic aspect of this interplay between atomic (or intrinsic) and hysteretic (or extrinsic) phenomena is that equilibration times vary from less than 1 ns to millions of years.

This determines, for example, the magnetic switching time of spin-electronics structures and the lifetime of information stored in magnetic recording media. This is related to the thermal instability of the magnetization direction known as superparamagnetism (Bean and Livingston 1959). This effect occurs in very small particles and is strongly temperature dependent.

In this book, the preferred *length unit* is 1 nm (10–9m), but 1 Å = 0.1 and 1 ìm = 1000 nm are also used, particularly for submicron features having sizes of several 100 nm. Recording densities are also measured in bytes per square inch or bytes per square centimeter (1 in –2 = 6.452 cm –2). Both SI and Gaussian units will be used for *magnetic quantities,* with explicit conversions occasionally included in square brackets. This also includes the parallel or alternate use of the SI unit tesla for coercivity and magnetization, as compared to the correct but cumbersome unit A/m.

The latter corresponds to *B* = ìo(*M* + *H*), and the former is obtained by incor porating ìo into *M* and *H,* so that one actually considers ìo *M* and ìo *H.* In the Gaussian system, multiplying the magnetization (emu/cm3) by the dimensionless number 4 changes the unit to gauss. (This is similar to measuring the perimeter of an island in miles and its diameter in feet.) Some everyday numerical *conversion rules* are: (i) 1 T = 10 kOe H" 0.8 MA/m (coercivity), (ii) 1 emu/cm3 = 1 kA/m and 1 T = 10 kG H" 0.8 MA/m (magnetization), (iii) 1 T 2= 100 MGOe H" 800 kJ/m3(energy product).

Finally, an SI susceptibility of 1 corresponds to a Gaussian susceptibility of 4. In the above con versions, we have used the numerical relation that 10/4 = 0.7958 H" 0.8.

Magnetic Nanostructures

Magnetic nanostructures can be produced in a wide range of geometries. Several chapters of this book deal with the fabrication, investigation, and application of individual geometries, such as nanowires and patterned thin films. The following paragraphs briefly characterize typical geometries, mention some systems of practical or scientific interest, and provide links to the individual chapters of this book.

Nanoparticles, Clusters, and Molecular Magnets

Small magnetic particles exist in nature or are produced artificially. Nanoparticles have sizes ranging from a few nanometers to submicron dimensions, whereas molecular magnets contain a few magnetic atoms in well-defined atomic environments. Clusters are inter mediate structures, with less well-defined atomic environments but ex hibiting atomic features such as facets. Examples of naturally occurring nanoparticles are magnetite (Fe_3O_4) nanoparticles precipitated in bacteria, insects and higher animals, and magnetite and other oxide particles responsible for rock magnetism.

Nanobio-magnetics is concerned not only with questions such as the role of magnetite particles for horizontal and vertical orientation of animals but also with important medical issues, such as local drug administration and cancer diagnosis. The small remanent magnetization of magnetic rocks, first analyzed by Neel in the 1940's, is exploited, for example, in archeo-magnetic dating and to monitor changes in the Earth's magnetic field. Small oxide particles, less than 10 nm in diameter, are observed in gels having the nominal composition FeO(OH) nH_2O. Fine particles are also encountered in meteorites.

Some artificially produced magnetic nanoparticle structures are Fe in Al_2O3 and so-called "elongated single-domain particles." Interesting applications of small particles are stable colloidal suspensions known *ferrofluids. A* variety of materials can be used, such as Fe_3O4, $BaFe_{12}O_{19}$, Fe, Co, and Ni, and a typical particle size is 10 nm. Most ferrofluids are based on hydrocarbons or other organic liquids, whereas water-based ferrofluids are more difficult to produce. They are used as liquids in bearings and to monitor magnetic fields and domain configurations.

Granular Nanostructures

Embedded clusters, granular materials, and other bulk nanostructures are of great importance in nanoscience. The structural correlation lengths of typical nanocomposite materials range from about 1 nm in x-ray amorphous structures to several 100 nanometers in submicron structures. Magnetic glasses and atomic-scale defect structures are beyond the scope of nanomagnetics, but they are of indirect interest as limiting cases and because nanomagnetic phenomena have their quantum-mechanical origin in atomic-scale magnetism.

Depending on grain size and microchemistry, granular nanostructures are used for example as permanent magnets (Nd-Fe-B), soft magnets (Fe-Cu-Nb-Si-B), and magnetoresistive materials (Co-Ag). There are two types of exchange-coupled permanent magnets: isotropic magnets, which exhibit random anisotropy and remanence enhancement, and oriented hard-soft composites, which utilize exchange coupling of a soft phase with a high magnetization to a hard skeleton. Closely related systems with many potential applications are magnetic clusters deposited in a matrix. For example, the narrow size distribution of 10–20% makes this material interesting as a granular media for magnetic recording. A well-known *soft-magnetic* nanocomposite is the "Yoshizawa" alloy Fe73.5Si13.5B9Cu1Nb3, which consists of DO3-structured FeSi grains embedded in an amorphous matrix.

Particle Arrays and Functional Components

Two-dimensional arrays of nanoparticles are of interest as scientific model systems and have many present or future applications. For example, advanced magnetic recording media can be characterized as a complex array of magnetic particles, and interest in dot arrays has been sparked by the search for ever-increasing storage densities in magnetic recording. In very small dots, quantum-mechanical effects are no longer negligible, and there are phenomena such as quantum-well states. These effects are of interest in quantum computing and spin electronics.

Most easily produced and investigated are submicron dots made from iron-series transition metals, such as Ni, but it is also possible to use metallic alloys, such as Permalloy, and to reduce the dot size to less than 100 nm. The dots may form square or hexagonal arrays, or structures such as corrals. Among the investigated phenomena are the properties of individual dots and interdot interactions. A related class of nanostructures are *antidots,* that is, holes in a film rather than dots on a film.

Potential applications include magnetic recording, sensors, magnetic and quantum computing, micron- and submicron-size mechanical devices, short-wavelength optics, and spin electronics. Other "functional" building blocks are, for example, nanojunctions, spin valves, and tips for magnetic-force microscopy (MFM tips).

Nanowires

There is a smooth transition from elongated dots and thin-film patches to nanowires. Magnetic nanowires have present or potential applications in many areas of advanced nanotechnology, including patterned magnetic media, magnetic devices, and materials for microwave technology. There a various methods to produce nanowires, such as deposition on vicinal surfaces and electrodeposition, including electrodeposition into porous alumina templates (Section 3). Much of the early work on magnetic nanowire arrays was concerned with exploratory issues, such as establishing an easy axis for typical preparation conditions, the essential involvement of shape anisotropy, as opposed to magnetocrystalline anisotropy, and the description of magnetostatic interactions between wires. More recently, attention has shifted towards the understanding of magnetization processes, such as the transition from curling-type to quasi-coherent nucleation, the influence of deposition-dependent polycrystallinity of typical transition-metal nanowires. Some other interesting phenomena are magnetic-mode localization, as evident *e.g.* from experimental activation volumes, spin-waves, and current-induced magnetization reversal.

MAGNETIC THIN FILMS AND MULTILAYERS

Magnetic thin films and multilayers can be classified as magnetic nano-structures, too, but it is common to treat homogeneous thin films and multilayers as a separate branch of magnetism, intermediate between nano-magnetism and surface magnetism. However, many recently developed and investigated nano-structures are thin-film nanostructures. Examples are self-assembled thin-film nanostructures, patterned nanomagnetic thin films, hard-magnetic thin-film nanostructures and thin films for magnetic recording.

Semihard thin films are used in magnetic recording media and have, more recently, attracted attention as tools for magnetic information pro cessing. In addition, on a length scale of a few interatomic distances, there is a variety of interesting thin-film effects, such as vicinal and interface anisotropies, moment modifications at surfaces and interfaces, thickness-dependent domain-wall and coercive phenomena, interlayer exchange-coupling, and finite-temperature magnetic ordering. A specific example is the nanoscale exchange-coupling or "exchange-spring" effects in multilayers.

Fabrication and Characterization

The broad variety of magnetic nanostructures corresponds to a diverse range of processing methods. The suitability of individual methods depends on the length scale and geometry of the nanostructures. In addition, each method is usually restricted to a relatively narrow class of magnetic materials. *Granular nanostructures* are produced by methods such as mechanical alloying and chemical reactions.

A traditional though somewhat cumber some method to fabricate *nanoscale particle arrays* of magnetic, dots, and wires on thin films is nanolithography. Other examples are molecular-beam epitaxy, the use of STMs, and chemical vapor deposition. The call for well-characterized large-area arrays of nanoparticles has stimulated the search for advanced production methods such as laser-interference lithography (LIL), where laser-intensity maxima effect a local transformation of a nonferromagnetic material into ferromagnetic islands. Another development is the use of focused ion-beam milling (FIB) to create small particles and particle arrays with well-de-fined properties.

Thin-film *nanowires* are comparatively easily obtained by depositing magnetic materials on vicinal surfaces and by exploiting structural anisotropies of the substrate. They can be produced with thicknesses down to one or two monolayers. Electrodeposition of magnetic materials into porous alumina may be used to produce regular wire arrays (see Sellmyer *et al.* 2001 and references therein).

Other ways of fabricating cylindrical nanowires include deposition into molecular sieves, track-etched polymer mem branes, and mica templates.

By electrodeposition into porous anodic alumina it is now possible to produce hexagonal Fe, Co, and Ni nanowire arrays with diameters ranging from 4 to 200 nm, and lengths of up to about 1 ìm, and variable center-to-center spacings of the order of 50 nm. The resulting materials are of interest as magnetic recording media, for optical and microwave applications, and as electroluminescent display devices. Aside from the above-mentioned iron-series transition-metal elements, there is interest in depositing alloys and multilayers, such as Fe/Pt, into porous templates.

The structural and magnetic characterization of magnetic nanostructures is the main focus of Ch. 5 and of various sections and subsections through out the book. Structural correlation lengths can be probed for example by X-ray diffraction, small-angle neutron scattering (SANS) and electron microscopy. Magnetic measurements are performed with the methods known from bulk and surface magnetism, although some techniques must be adjusted to the small signals from certain structures. Examples are vibrating sample magnetometry (VSM), magneto-optical Kerr effect (MOKE) measurements, and SQUID magnetometry. Some methods, such as magnetic-force magnetometry, are nanospecific and presently being applied to measure hysteresis loops of nanoscale magnetic particles.

Applications

Magnetic nanostructures are used in the form of traditional magnetic materials, such as hard and soft magnets, and in specific functional structures, such as sensors. Hard or *permanent magnets* are used, for example, in electromotors, hard-disk drives, loudspeakers, windshield wipers, locks, refrigerator magnets, and microphones. Some applications, such as toys, do not usually require high-performance magnets, but hard-disk drives and other high-tech applications require highly sophisticated rare-earth permanent magnets with well-defined nanostructures. Compared to the highly anisotropic hard magnets, *soft magnets* exhibit very low magnetic anisotropy. They are widely used for flux guidance in permanent-magnet and other systems, in transformer cores, and for high-frequency and microwave applications, and in recording heads. In advanced soft-magnetic materials, nanostructuring is used to reduce magnetic losses by controlling anisotropy, eddy-current losses, and other properties. A key application and driving force of magnetic nanotechnology is *magnetic recording media.* They are used not only for audio-visual technology, for example in audio and video tapes, but also in computer technology, for example in hard-disks.

Aremarkable increase in areal density of many orders of magnitude in the last two decades has relied heavily on nanostructuring of media and read and write heads. Artificial nanostructuring is a way of creating completely new technologies. One area is spin electronics, and various types of nanostructures, such as multilayers and nanojunctions, are being used or investigated in this context. The magnetoresistance of metallic thin films, granular nanostructures and magnetic oxides are exploited in sensors, and a problem of current interest is spin injection into nonferromagnetic metals and magnetic semiconductors. Other recent developments are magnetic nanostructures for quantum computing, multiferroics (where nanoscale effects are exploited to synergize electric and magnetic degrees of freedom), and nanoparticle ferrofluids for cancer treatment, guided by a magnet and delivering high local doses of drugs or radiation. Nanoscale effects are also exploited in micro-electromechanical systems (MEMS) and magnetic-force nanotips made from CoPt.

THE MAGIC OF MAGNETIC MULTILAYERS

In recent years there has been a renaissance in the study of magnetic multilayers—primarily because of the discoveries of enhanced or giant magnetoresistance and oscillatory interlayer exchange coupling in magnetic multilayered structures consisting of alternating thin ferromagnetic and nonferromagnetic metallic layers. These discoveries have led to intense theoretical and experimental studies of an extraordinary range of structures over the past few years.

Giant magnetoresistance (GMR) is displayed by a wide variety of inhomogeneous magnetic nanostructures consisting of magnetic entities, for example, thin layers or particles, separated by thin nonferromagnetic metallic spacer layers. The original observation of GMR was made in single-crystalline (100) Fe/Cr "sandwich" and superlattice structures grown by molecular beam epitaxy (MBE) in ultrahigh-vacuum deposition chambers.

It had previously been found that the magnetic moments of the Fe layers in a (100)-oriented Fe/Cr/Fe sandwich are aligned antiparallel to one another in zero field when the Cr layer is about 9 Å thick. When a sufficiently large magnetic field is applied, the moments of the Fe layers become aligned parallel with one another. The resistance of such sandwiches or superlattices comprising many such sandwiches depends upon the magnetic arrangement of the magnetic layers and was observed to be higher when the moments are aligned antiparallel to one another. The magnetoresistance of such structures is much larger than that of the intrinsic magnetoresistance of the Fe layers themselves and is especially large at low temperatures.

While the observation initially appeared to be somewhat esoteric, it was found soon afterward that similar results could be obtained in polycrystalline Fe/Cr structures grown by the much simpler technique of magnetron sputtering. These experiments also revealed that GMR could be observed in a wide variety of transition-metal magnetic multilayers. In addition, the experiments showed, in contrast to the earlier work on MBE-grown structures, that the magnitude of the GMR effect oscillated as the thickness of the nonferromagnetic spacer layers between the ferromagnetic layers was increased.

This oscillation was shown to be caused by an oscillation in the sign of the interlayer exchange coupling between the ferromagnetic layers. The coupling was shown to oscillate between antiferromagnetic and ferromagnetic coupling such that the magnetic moments of successive ferromagnetic layers were either parallel (ferromagnetic) or antiparallel (antiferromagnetic) in small magnetic fields. Oscillatory coupling was shown to be a very general property of almost all transition-metal magnetic multilayered systems in which the nonferromagnetic layer comprises one of the *3d*, *4d*, or *5d* transition metals or one of the noble metals. The oscillation period was found to be just a few atomic layers, typically about 10 Å, but varying up to ~20 Å. Only those multilayers for which the interlayer coupling is antiferromagnetic display significant giant magnetoresistance effects. It is only in those systems that the relative orientation of the magnetic moments of neighboring layers is significantly altered by applying a magnetic field.

Magnetron sputtering allowed the study of many different magnetic multilayered systems. One of these systems, namely ferromagnetic cobalt layers separated by thin copper layers, was found to exhibit very large GMR effects even at room temperature. Values of GMR in Co/Cu multilayers exceed 110% at room temperature[2]. While the largest GMR values require magnetic fields exceeding ~20 kOe, magnetoresistance values of 50-60% are obtained in fields of several hundred oersteds and values of ~20% or more in fields of a few tens of oersteds.

These lower values are obtained by using thicker Cu layers, for which the interlayer exchange coupling is weaker. Such large magnetoresistance values at room temperature make such multilayers attractive candidates for a variety of technological applications. Moreover, the fact that these multilayers are readily formed using deposition techniques compatible with large-scale manufacturing makes them even more alluring.

Magnetoresistive materials are used in a wide variety of applications for detecting magnetic fields. For some applications, such as the detection of the rotation of a spinning object, as for example in anti-lock brake systems in automobiles, relatively large magnetic fields can be used.

For other applications, such as the detection of magnetic bits in a hard disk drive, materials whose resistance is sensitive to quite small magnetic fields are needed. Conventional materials such as Ni-Fe alloys exhibit resistance changes at room temperature of just a few percent in magnetic fields of a few oersteds. GMR sandwiches can achieve sensitivities to such fields of perhaps as much as five times greater than conventional materials.

GMR multilayered structures have already found their way into leading-edge hard disk drive products. Early in November 1997, IBM was the first company to announce

GMR magnetic recording read heads in a family of disk drive products, designated Deskstar* 16GP. First customer shipment began in January 1998. In these products, the use of GMR read heads allows the reading of extremely small magnetic bits at an areal density of 2.69 gigabits per square inch.

The capacity of a hard disk drive is largely determined by the areal density in conjunction with the size and number of disks or platters within the drive. GMR read heads allow more than 3.2 gigabytes of data to be stored on each 95-mm-diameter disk of the Deskstar 16GP disk drives.

The resulting data storage capacity of these "3.5-inch form-factor" drives is 16.8 gigabytes. The areal density of hard disk drives, driven by the rapid evolution of the various technologies involved, is increasing at a compound growth rate of about 60% per year, and by the year 2001 is expected to reach 10 gigabits per square inch.

This issue of the *IBM Journal of Research and Development* discusses some of the work carried out over the past few years, largely within IBM, concerning GMR, oscillatory interlayer exchange coupling in magnetic multilayers, and related effects.

The first paper, by Allenspach and Weber, discusses some experiments which probe the origin of the oscillatory interlayer coupling in single-crystal (100)-oriented Co/Cu/Co sandwiches prepared on single-crystal Cu substrates. These experiments have shown that the strength and period of the oscillatory coupling could largely be explained from the well-known electronic properties of Cu. In these experiments, the thickness of the Cu and Co layers could be varied by fractions of an atomic layer, and it was observed that the magnetic properties of this system depend on such minute variations.

The second paper, by Jones, discusses some of the different theoretical approaches used to understand oscillatory interlayer coupling. In particular, Jones discusses the role of quantized electronic states in magnetic multilayers and their dependence on the magnetic structure of the multilayer.

In the following paper, Himpsel et al. discuss the experimental observation of such quantum-well states in (100) Co/Cu using spin-polarized photoemission spectroscopy and the relationship of such states to both interlayer coupling and GMR.

While sputter-deposition techniques account for many of the major developments in the field of magnetic multilayers in recent years, MBE continues to be a valuable technique for the growth of certain single-crystalline systems.

Farrow discusses the role MBE has played in developing an understanding of certain aspects of GMR and oscillatory interlayer coupling. Much work was applied to understanding why single-crystalline (111)-oriented Co/Cu multilayers grown by MBE show no GMR, whereas polycrystalline (111)-textured Co/Cu multilayers grown by sputter deposition show very large GMR values. For many years this was a puzzle which was addressed by many groups. It is now generally believed that imperfections in the MBE-prepared multilayers account for this behavior.

The use of advanced materials growth techniques, such as the use of surfactants to promote smoother films, has finally led to the observation of oscillatory interlayer coupling in MBE-grown (111) Co/Cu. These studies emphasize that subtle structural modifications of magnetic multilayers can lead to dramatically altered properties.

Nesbet reviews theories of GMR and, in particular, discusses models of GMR based on realistic band structures of the metals involved.

These models demonstrate the importance of the scattering at the interfaces between the ferromagnetic and nonferromagnetic layers in GMR. The importance of interface scattering to GMR, originally demonstrated in experiments which modified the interfaces, is now widely accepted and is of particular technological significance.

In recent years there have been many advances in experimental techniques for probing aspects of the structure and magnetism of magnetic layers and multilayers. Stöhr and Nakajima discuss the use of advanced synchrotron-based techniques to probe, with elemental specificity, not only the magnetic moments of the ferromagnetic layers in magnetic multilayers but also the magnetism of the nonferromagnetic layers.

Copper layers in Co/Cu multilayers become magnetically active by virtue of their proximity to the ferromagnetic Co layers. By contrast, in Fe/Cu structures, the ferromagnetic Fe layers can become magnetically inactive because of the structural influence of the Cu layers.

In the paper which follows, Sun et al. discuss recent developments in an entirely different class of materials: the manganite perovskites. These materials show extremely large changes in resistance in magnetic fields, but typically at temperatures well below room temperature and in extremely large magnetic fields.

The origin of their magnetoresistance, commonly referred to as colossal magnetoresistance, is quite different from the GMR of magnetic multilayers. In my opinion, it is unlikely that such materials will be useful technologically because of their great complexity as well as the strong temperature dependence of their properties. In contrast to these materials, GMR multilayered structures have already found important technological applications.

In the final paper of the group, Tsang et al. discuss the use of specially engineered GMR structures for hard disk drive read-head applications. As mentioned above, GMR multilayered structures can display very large magnetoresistance values, but at relatively large magnetic fields.

In a magnetic recording disk drive, information is stored as magnetized regions within thin magnetic layers (which themselves can be rather complicated multilayered thin-film structures) deposited onto aluminum or glass platters.

The regions are magnetized longitudinally within the plane of the platter either parallel or antiparallel to the circumference of the platter. It is the transitions between these magnetized regions which are detected by a magnetic recording read head via the magnetic flux emanating from the transition regions. The strength of the magnetic field at the read head is only a few tens of oersteds. Thus, a field-sensing device capable of detecting such small magnetic fields is required.

GMR magnetic multilayered structures can be engineered to be sensitive to such small fields with greater sensitivity than conventional ferromagnetic metals. For example, this can be achieved by forming simple sandwiches of two ferromagnetic metals separated by a thin layer of Cu perhaps 20 Å thick.

One of the ferromagnetic layers is magnetically hardened so that for the range of magnetic fields to which it is subjected, the magnetic moment of that layer is fixed. This can be achieved by coupling that ferromagnetic layer to a thin antiferromagnetic layer using the phenomenon of exchange biasing. In addition, one takes advantage of the fact that whereas the GMR effect falls off very slowly with increasing separation of the ferromagnetic layers for Cu spacer layers, the magnitude of the interlayer exchange coupling decreases much more rapidly with increasing Cu thickness. Thus, the intrinsic magnetic coupling through Cu layers 20 Å thick is weak.

The moment of the second ferromagnetic layer in the GMR sandwich rotates in the presence of the magnetic field from the transition region and thus causes a change in the resistance of the sandwich. The variation in the resistance of the sandwich is used to detect the transition region between the magnetic bits, and thus the stored data can be recovered.

This type of GMR sandwich has been termed a spin-valve GMR sandwich. Tsang et al. discuss the optimization of GMR multilayered structures for high-density magnetic recording read-head applications. For such applications, very thin ferromagnetic sense layers are needed. One of the main reasons why GMR structures provide a greater signal for these applications than conventional ferromagnetic alloys is the predominant interfacial origin of GMR, which typically results in a larger GMR signal for thinner ferromagnetic layers.

The papers included in the GMR-related portion of this issue cover just a small subset of the work carried out within IBM over the past few years on magnetic multilayers and related structures. This work is itself just a small subset of that carried out within the scientific community. It is perhaps remarkable that within less than a decade, magnetic multilayers have evolved from a scientific curiosity to become materials of significant technological importance.

This class of magnetic multilayers has revealed a considerable number of scientific insights. In addition, the study of its properties has led to significant developments in the preparation of magnetic multilayered structures with layers just two or three atomic layers thick. Such structures can now be prepared routinely and with sufficient reproducibility for commercial applications. Advances in structural, magnetic, and electronic characterization have helped to motivate continuing interest in such structures. In the future, studies seem likely to evolve to the study of lower-dimensional magnetic nanostructured materials patterned or structured into lines or dots.

2

Magnetic Nanostructures: From Fundamental Studies to Technical Applications

Magnetic nanostructured materials produced in the present are reviewed. Some of their technical applications are presented.

Introduction

The place of scientific research in economical and social life is of great importance. The generated knowledges and their interactions with industry contributed essentially to the evolution of the human society. Since our summer school is mainly connected with the field of magnetism, the evolution of the research in this field and the connection with nanomaterials seem to be useful to be remembered.

The development of the knowledge in the field of magnetism may be divided in seven periods: (1) ancient (-1000 | 1500), (2) early (1500 | 1820), electromagnetic (1820 | 1900), understanding (1900 | 1935), high frequency (1935 | 1960), applications (1960 | 1995) and spin electronics (1995 ÷ present). Atomic scale magnetic phenomena were discovered in the first half of the last century, but only in recent decades it became clear that solid-state magnetism is, to a large extent, a nanostructural phenomena. A naturally occurring biomagnetic phenomenon is magnetite nanoparticles precipitated in bacteria, mollusks and contain chains of 40-100 nm particles used for vertical orientation. Similar magnetite particles have been found in the brains of bees or pigeons and may serve as field sensors for migration.

The recent improvement of magnetic materials knowledge is connected to their nanostructuring. The artificial nanostructuring create completely new materials and technologies.

The progressing miniaturization in computer technology is based mainly on nanostructured media for ultra-high density magnetic recording. Related to the above is the spin electronics and various types of nanostructures such as multilayers and nanojunctions. Another area is constituted by magnetoresistive sensors based on magnetoresistance effect in metallic thin films, granular systems and magnetic oxides. The nanoparticles can be used also in medicine, micro- electromechanical systems, various nanodevices as nanoscale magnetic-force nanotips or permanent magnets.

Magnetic recording technology has an excellent future growth potential. The requirements connected with a reduction in the peripheral device size and power consumption while maintaining an exponential increase in storage capacity is strongly related to the development of nanomaterials. The various types of storage in relation capacity-average access time is given in Fig.1a and in Fig.1b a hierarchy based on access time and cost per bit is shown.

Although the access rate in semiconductor memories is fast and capacity is high, this type of memory is still very expensive and in addition the memory is volatile. The disk drives offer longer access time but they are cheaper and are used in standard PC and laptop applications. The magnetic memories are also nonvolatile. Magneto-optical memories now grow as importance since of their good average access time and low price.

Nanostructure Materials

The magnetic nanostructured materials may be characterized by diversity of geometries. Various types of small magnetic particles exist in nature or were artificially manufactured. A somewhat arbitrary division may be as:

(1) particle and cluster,

(2) thin films and multilayers,

(3) particle arrays,

(4) nanowires,

(5) nanocomposites.

There are chain of fine particles (a), striped nanowires (b), cylindrical nanowires (c), nanojunctions (d), vicinal surface step (e), nanodots (f), antidots (g) and particulate medium (h). The elongated single domain (ESD) particles are commonly used for magnetic recording. In addition, small particles can be used to produce ferrofluids. In this case typical dimensions of particles are 10 nm and are realized from Fe_3O_4, $BaFe_{12}O_{19}$, Fe, Ni etc.

The nanoparticles may be also arranged in both free- and embedded clusters. Two-dimensional arrays of nanoparticles are useful in advanced recording media. The dots disposed on a thin film are of submicron dimensions and are made from 3d series transition metals or 3d magnetic alloys. The geometry of dot arrays may be square, hexagonal or to form other structures as corrals. The antidots nanostructures are constituted as holes in the film and through potential applications are: magnetic recording, sensors, mechanical devices, spin electronics. The magnetic thin films and multilayers can be considered as nanostructures. Since of their use in magnetic recording devices, the thin films were largely studied. Magnetic nanowires have potential applications in advanced nanotechnology including patterned magnetic media, magnetic devices and as materials for microwave applications.

They can be produced with thickness down to one or two monolayers. By electrodeposition may be also obtained a regular wire arrays. Cylindrical nanowires may be obtained be deposition into molecular sieve. By electrodeposition into porous anodic alumina was possible to produce Fe, Co and Ni wires with diameter d = 4 - 200 nm and length l E1 ∝m. Typically, the nanowires form nearly hexagonal columnar arrays with centre to centre spacing of E" 50 nm. These systems are of great interest for magnetic recording media.

MAGNETIC NANOMATERIALS IN DATA STORAGE APPLICATIONS

Nanomagnetic systems are an important part of today's technology and have potential applications in future new devices. The applications of nanomaterials as patterned media for hard disks, miniature magnetic sensors using GMR effect and magnetic memory cells which are being developed for fast and dense solid state memory will be outlined. The areal density of magnetic recording has been increased over the past few years at 60 % per annum due to concurrent optimization of successive generations of thin film heads, miniaturization of the planar write head and improvement of Co-based in-plane thin film media.

The hard disks having a greater storage density have been developed. However, the effects associated with the finite grain size of magnetic alloys used for storage, limit the miniaturization. As a result a new type of storage media has been proposed made from a closely-packed array of nanomagnets patterned onto a hard disk. A conventional hard disk is constituted by a thin film of granular Co-alloys. The thin films are constituted from exchange decoupled grains, each grain reversing individually their magnetization. Intergranular exchange coupling is suppressed by adding a non-magnetic element (particularly Cr) which segregates at the grain boundaries.

In order to increase the storage capacity, it is necessary to reduce the size of the grains and to ensure a minimum number of grains per bit, to limit statistical noise. It was suggested that for obtaining low particle noise contribution, about 1000 grains are needed in one bit cell. For example a bit cell in the Co-based medium used for the IBM (0.184 Gbit cm-2) consists of about 1900 grains with a grain size of about 20 nm. Thus, for a 1.5 Gbitcm-2 thin film medium, a grain size of about 10 nm is necessary. In this case there will be about 600 grains in a 1 bit area.

The reduction of the size of grain is limited to E" 10 nm, since the grains having smaller dimensions are superparamagnetic. Smaller grains can only be used if their switching field is increased. This may be possible by increasing the coercivity of Co alloys, up to 0.28 MAm-1, by incorporating Ta, Pt. Coercivities up to 0.8 MAm-1 would be possible by using permanent magnet materials (Nd-Fe-B, Sm-Co). In this case the grain size can be reduced up to 2.5 nm but in this case it is not possible to use write heads, available today, in which the write fields are limited to maximum 0.4 Mam-1.

As mentioned already, higher storage densities, in the range (16 – 80) Gbitcm-1, could be reached by using patterned media. The collection of exchange coupled grains in this case is replaced by an array of nanomagnet bits (10 – 20 nm sizes) physically isolated from their neighbours and able to be switched "independently".

The bit could be oriented either in the plane of disk or perpendicular to it. In Fig. 3 there are plotted three types of patterned media with bars in the plane of the disk (a), perpendicular pillars (b) and cones (c). The perpendicular configuration leads to a more dense packing, but in-plane configuration allows a more conventional reading and writing technology to be used. Nanomagnet bars of Co film with dimensions 40x25x200 (nm), in plane, have a storage density of 4 Gbitcm-2. The average switching field is E" 0.1 MAm-1. We note that the packing density is limited by the magnetostatic interactions between elements.

Giant Magnetoresistance Sensors

Magnetic memory elements are being developed as a new type of non-volatile random access memory (RAM). As mentioned before, in certain applications magnetic RAM could replace silicon devices based on CMOS technology, particularly where the resistance to radiation is important. The operation of magnetic memory cells is based on the giant magnetoresistance effect (GMR) in spin valves or spin tunnel junction structures and special properties of nanomagnets as storage elements for binary data. Giant magnetoresistance was first discovered in Fe-Cr superlattices with the non-magnetic Cr layers less than 2nm thick, separating slightly thicker Fe layers.

The origin of the GMR effect is the difference in scattering of electrons of opposite spins (up or down) as they pass through a ferromagnetic materials. The GMR sensor could consist of two magnetic layers which have their magnetizations oriented parallel or antiparallel to each other. When the two layers are magnetized antiparallel, the spin-up and spin-down electrons will be scattered equally. When the layers are magnetized parallel, the spin-dependent scattering results in a lower resistivity for one type of electrons and a higher resistivity for the other type. The lowering resistivity for one type, result in a lower resistance for the structure, as a whole device. Thus, when the direction of magnetization direction is changed, multilayers structures, spin valves and spin tunnel junctions can be used as sensors.

This involves that they must be patterned into micron or sub-micron sized elements. Multilayers comprising of alternating series of magnetic - non-magnetic bilayers can produce large "R/R values, but requires a field of the order 10-40 kAm-1 to reverse the antiparallel coupling of layers and to saturate them in the low resistance state. Thus, the multilayers cannot be used as detectors of small fields. A more sensitive response is possible by using a spin valve structure. The spin valves have two soft magnetic layers (Ni-Fe) separated by a metallic spacer layer (Cu).

The GMR effect is generated by having a "pinned" layer held rigidly to a fixed magnetization direction and a layer in which the magnetization is free to rotate relative to the pinned layer when a field is applied. Thus, high resistance (antiparallel) and low resistance (parallel) states can be achieved. Spin tunnel junctions have two magnetic electrodes separated by an insulating tunnel barrier made from aluminium oxide. Unlike in spin valves, the current flows through the junction perpendicular to the layers. This increases the effect of spin dependent scattering on the electrons passing through the structure and leads to higher "R/R values than in spin valves. The interesting physical properties of magnetic nanostructures materials as well as their characterization will be largely presented in the following lectures. The analyze methods will be also described. In addition come of their technical applications will be outlined.

Magnetism is the underlying effect of many technical achievements provided by science and industry during the last decades. In today's research focusing on nanotechnology and electronic data storage, interest in magnetism even increased, and future attempts in micro-design require a deeper knowledge of magnetic properties on the nanoscale. In contrast to optical or acoustic phenomena, which in principle are accessible by human perception, magnetism cannot be directly sensed and its experimental investigation is restricted to indirect observation. This can be done by measuring the effect of magnetic order on other physical properties of a solid like electrical and thermal conductivity, elasticity, heat capacity, sound attenuation or lattice strains – all these properties show anomalies at the magnetic phase transitions.

Another important access to magnetism is given by magneto-optical (MO) effects, which relate the optical properties and the magnetic state of a solid. In the visible light region MO effects are widely used to measure the average magnetization of thin samples (thickness << wavelength of the probing light) and therefore has become a standard tool in thin- film magnetism research. Although MO effects in the visible-light region are quite small, refined optical methods not only yield sufficient contrast for distinguishing bits of opposite magnetization, but even allow one to observe details of magnetic domains in optical microscopes. Most importantly, MO techniques are ideally suited for studying magnetization reversal processes in external magnetic fields, which is impossible with, e.g., electron microscopes involving slow cascade electrons that are strongly affected by the Lorentz force.

Besides all merits, conventional MO techniques lack element specificity, simply because electric dipole transitions in the visible-light region mainly take place between delocalized conduction-electron band states. This has become a severe limitation in analyzing compounds containing several magnetic elements, as required for information storage or permanent-magnet nanostructures. In these advanced magnetic systems, lanthanide elements (mostly Tb) are used to achieve large perpendicular magnetic anisotropy, and large coercive fields.

Reliable element-specific information can be obtained by techniques that involve core-level transitions, e.g. by magnetic circular dichroism in x-ray absorption (XA). In the soft x-ray region, absorption signals are commonly detected by electron-yield methods, so that in general no large external magnetic fields can be applied, and hence only very few hysteresis loops have been recorded via electron yield. Alternative XA detection modes which do not employ electrons are either limited to specific samples that are suspended on transparent foils, or they suffer from saturation effects for all but highly diluted samples (fluorescence yield).

Thus, to probe the field-induced magnetization reversal of a compound in an element-specific way, it is useful to employ x-rays in *reflection* like in conventional magnetooptics.

Resonant enhancement of x-ray scattering cross sections at inner-shell absorption thresholds have been used for years to investigate the magnetic structure of lanthanide and actinide materials in the hard x-ray region (above 2 keV). The experimental demonstration of large changes in the specularly reflected x-ray intensity at the Fe L 2,3 threshold upon magnetization reversal initiated the ongoing search for magneto-optical (MO) effects in the *soft x-ray* region, and for ways to apply them to element-specific studies of heteromagnetic systems.

Apart from their technological relevance in soft and hard magnetic films, lanthanide elements constitute a unique case in x-ray optics: the *4d* “ *4f* electric dipole transitions (the so-called giant resonances) are among the strongest in the periodic table. They give rise to high absorptivity and reflectivity, corresponding to a very short x-ray absorption length which in the course of the present work were found to be comparable or even shorter than the lanthanide *4d* “ *4f* transition wavelength (*l H*” 8 nm).

This unique situation is quite different from previously studied transition-metal (Fe, Co,...) *2p* “ *3d* transitions, where the attenuation lengths are much longer than the wavelength, giving rise to complicated interferences when samples are structured on the nanometer scale. Yet, in analyzing soft x-ray MO signals from thin films and multilayer systems with thicknesses comparable to the x-ray wavelength, previous investigations have shown that a comparison with model calculations (based on the Fresnel equations) is needed in order to extract layer-resolved sample magnetization profiles. Several experimental determinations of soft x-ray MO constants have been reported for ferromagnetic transition metals in the region of the L2,3 thresholds, but none so far for the lanthanide elements.

The purpose of this thesis is to investigate MO effects of lanthanide elements in the soft-x-ray region. The results are compared with visible-light MO data in order to discuss their mutual compatibility. Special attention is paid to the element-specific behavior of heteromagnetic systems. The elements Gd and Tb are chosen as examples because of their similarity in lattice constants but widely different magnetic behavior.

SCALABLE PARALLEL MICROMAGNETIC SOLVERS FOR MAGNETIC NANOSTRUCTURES

A parallel finite element micromagnetics package has been implemented, that is highly scalable, easily portable and combines different solvers for the micromagnetic equations. The implementation is based on the standard Galerkin discretization on tetrahedral meshes with linear basis functions. A static energy minimization, a dynamic time integration, and the nudged elastic band method have been implemented. The details of the implementation and some aspects of the optimization are discussed and timing and speedup results are given. Nucleation and magnetization reversal processes in permalloy nanodots are investigated with this micromagnetics package.

In various fields of computer aided engineering like structural analysis, fluid dynamics, and electromagnetic field computation, as well as micromagnetics the finite element method has been successfully applied. In particular, its flexibility in modeling arbitrary geometries has made it very popular. In the light of the importance of the microstructure of magnetic materials the finite element method has been chosen for the implementation of a micromagnetics package.

There are several commercial and open source micromagnetics packages available, however all of them use the finite difference method. Our finite element micromagnetics package combines several unique features:

It is:

- entirely based on portable, free, open source software packages,
- highly portable to different hardware platforms, which range from simple PCs to massively parallel supercomputers,

- highly optimized and scalable,
- well integrated, combining static energy minimization, dynamic time integration, and nudged elastic band methods.

FINITE ELEMENT MICROMAGNETICS

The finite element method requires the discretization of the spatial domain &! with "finite elements" – a regular triangulation. For two dimensional problems triangles and rectangles, in three dimensions tetrahedral and hexahedral elements are commonly used. Also a mixture of different types of elements

is possible, but after the evaluation of various other implementations and for simplicity a tetrahedral discretization has been implemented. As compared to finite difference methods the finite element mesh may be entirely unstructured, which makes the modeling of complicated geometries and irregular microstructures more convenient. However, the creation of the geometrical model and its triangulation are still very demanding tasks, which require sophisticated (commercial) tools. Patran by MSC.Software and GID (there is a free academic version available) are suitable software packages for which import filters have been implemented. The required resolution of the finite element mesh – the maximum size of the finite elements – is determined by the smallest features which might occur in the solution of the PDE. These features have to be properly resolved and the approximation of the exact solution by the test functions has to be sufficiently accurate to give meaningful results.

In micromagnetics these limits are defined by three typical lengths:

*l*bex="*A*/*K*1, *δ*BW= *πl*bex(1)*δ*BW is the width of Bloch type magnetic domain walls (typically found in bulk hard magnets), A is the exchange constant and $K1$ is the magnetocrystalline anisotropy constant.

$l$nex="2μ0A/J2s, δNW= πlnex(2)δNW is the width of Neel walls (typically found in soft magnetic materials and thin films), Js is the saturation polarization. $l$thex="Ajs Hth, Hth="2αkBT'$t\gamma J$sl3(3)$l$thex is the thermal exchange length with the thermal field Hth, which depends on the Gilbert damping constant α, the temperature T, time step of the numerical time integration scheme "t, the gyromagnetic ratio γ, and the spatial correlation length l (which equals the cell size). In the most general case, the maximum cell size of the finite element mesh has to be smaller than the minimum of the three lengths defined above. However, if the structure of the solution is roughly known, it is possible to use an "adapted" mesh for the simulations. It has a high resolution (small elements) in areas with very small features (e.g. domain walls) and a low resolution (large elements) in other areas where the solution (magnetization) is very uniform. As a result, the number of nodes (and therefore the number of unknowns) and, thus, the time required for the simulation can be greatly reduced.

For dynamic time dependent problems a rigid adapted mesh is often not suitable, because the solution changes over time and the smallest features of the solution move through the mesh. If they leave the high resolution mesh and reach areas with larger finite elements, artificial pinning on the mesh will occur and give wrong results. Thus, an adaptive mesh refinement method is required, which changes the structure of the mesh during the simulation and adapts the resolution of the mesh to the solution. It can be shown, that these methods lead to (almost) optimal complexity and give most accurate results with the smallest numerical (computational) effort. Adaptive mesh refinement methods are still a very active research area, and they have been successfully applied also in numerical micromagnetics. However, this adaptive meshing technique has not been implemented (yet) in our micromagnetics package.

GIBBS FREE ENERGY

The total energy of a micromagnetic system is given by the Gibbs free energy *E*tot, which depends on the magnetic polarization, the external field and some (temperature dependent) material parameters. It includes macroscopic contributions such as the Zeeman energy and the magnetostatic energy as well as microscopic contributions like the magnetocrystalline anisotropy energy and the exchange energy. The external field is independent of the magnetization distribution and the exchange and anisotropy energy are short range interactions, which depend only on the local magnetization distribution.

Thus, they can be computed very efficiently. However, the magnetostatic field is a long-range interaction, that is the most expensive part in terms of memory requirements and computation time. Its calculation is usually based on a magnetic vector or scalar potential (cf. Sec. 2.3.4). In addition, it is an open boundary problem, for which various methods have been developed. The total Gibbs free energy is given by *E*tot=+"&!(*w*exch+ *w*ani+ *w*ext+ *w*demag) *dv* =(4)=+"&!(*A*(("*ux*)2+ ("*uy*)2+ ("*uz*)2)+*K*1(1 " (*a* · *u*)2) " *J* · *H*ext"12*J* · *H*demag)*dv*,where *J*(*x*,*t*) = *J*s(*x*) · *u*(*x*,*t*), |*u*| = 1(5)describes the magnetic polarization as a function of space and time.

A is the exchange constant, *K*1 is the first magnetocrystalline anisotropy constant and *a* the unit vector parallel to the easy axis, *H*ext the external field, and *H*demag the demagnetizing field.In thermodynamic equilibrium a micromagnetic system tries to reach a state with minimum total energy. The aim of micromagnetic theory is to find the magnetic polarization in equilibrium. Brown proposed a variational method, that is based on the calculation of the variational derivative of the total energy with respect to the magnetic polarization. In equilibrium (in an energy minimum) the coefficients of the linear term vanish for any variation $\delta u \delta E$totδu= 0.

This leads to Brown's equations u ×(2A"u + 2K1a (u · a) + JsHext+ JsHdemag)= 0(7) Thus, in equilibrium the magnetic polarization J is parallel to an "effective field"Heff = 2AJs"u +2K1Jsa(u · a) + Hext+ Hdemag(8) and the torque, which acts on the polarization, vanishes J × Heff= 0.

Since any contribution parallel to the polarization J does not add to the torque, it does not make any difference if the magnetic field H or the magnetic induction $B = \mu 0H$ + J is used for the effective field.

Gilbert Equation of Motion

The minimization of Eq. (4) can find an equilibrium magnetization distribution. However, the energy landscape of micromagnetic systems is usually very complicated and contains many local maxima, minima, and saddle points. Therefore, the choice of the initial magnetization distribution has a strong influence on the result. A more physical approach to the problem and a more realistic approach of the system to its equilibrium in a local minimum is provided by a dynamic description of the path through the energy landscape.

The motion of a magnetic moment in a magnetic field is mainly governed by its Larmor precession around the local magnetic field. The damping of the precession causes the relaxation to equilibrium. There are many processes that contribute to the damping in a magnetic solid like magnon-magnon and magnon-phonon interactions,

interactions between localized and itinerant electrons and eddy currents, for example.

The *Gilbert equation* describes the precession and combines all damping effects in a phenomenological damping term with a single damping constant *αdJdt*= –*γJ* × *H* + *αJ*s*J* ×*dJdt*,(10) where *γ* = 2.210173 × 105 mAs is the gyromagnetic ratio.

This formulation is equivalent to the *Landau-Lifshitz-Gilbert (LLG) equation dJ dt* = –*γ J* × *H* –*αγ J* s*J* × (*J* × *H*), (11) with *γ* = *γ*1+*α*2. The intrinsic timescale is determined by the Larmor frequency *ω* = *γH* eff, which is usually in the order of GHz. Thus, the precession time is smaller than a nanosecond, which requires time steps in the order of picoseconds or even less. This limits the maximum simulated time to about 100 ns.

EXCHANGE ENERGY

Since we know the external field explicitly, we can just simply add it to the other contributions to the effective field.

Demagnetizing Field and Magnetostatic Energy

The demagnetizing field is a little more complicated to handle, because it is an "open boundary problem" with one of its boundary conditions at infinity. The magnetostatic potential can be calculated directly by integration using the Green's function, but this method is very slow and expensive in storage. Fast multipole methods have been developed to overcome these problems. Fredkin and Koehler proposed a hybrid finite element/boundary element method that requires no finite elements outside the magnetic domain Ω. A comparison of these methods can be found in.

Since the matrices for the exchange and anisotropy energies depend only on the (time independent) material parameters and the geometry, they need to be calculated only once at the beginning. In order to save time and memory, they can also be assembled into a single combined matrix, if the energies and fields are not required separately.

SOLUTION OF THE MICROMAGNETIC EQUATIONS

Energy Minimization

In Sec. 2 we have introduced the Gibbs free energy of a micromagnetic system, discretized various contributions to the total energy, and derived the matrix-vector formulation. Some complication was introduced by the demagnetizing field, but the hybrid finite-element/boundary element method provides an elegant way to solve the problem accurately with a finite element mesh that is restricted to the magnetic bodies. This enables us to implement a simple energy minimization scheme to find the equilibrium magnetization distribution.

However, it has been shown that the use of spherical coordinates has several advantages:

- The number of unknowns is reduced by 1/3, which speeds up the solver.
- The norm is automatically preserved, which allows the use of an unconstrained solver.
- Convergence problems with a Lagrange multiplier approach for the constraint $|J| = J$s are avoided.

However, the direct calculation of the energy gradient in spherical coordinates causes various problems due to the periodicity of the polar and azimuth angles.

Therefore, the magnetic polarization for the minimizer is given in spherical coordinates. Then it is converted to Cartesian coordinates. The energy gradient is calculated in Cartesian coordinates, converted back to spherical coordinates and returned to the minimizer.

For the minimizer itself the limited memory variable metric (LMVM) algorithm – a quasi-Newton-method – of the TAO package has been selected, because it requires only the function values and the gradient of the total energy. In replacement for the Hessian (which is not available due to the demagnetizing field) the second-order information is approximated by a limited history of previous points and gradients. A similar method has been used in and showed better convergence rates than Newton- or Gauß-Seidel methods.

The Dynamic Equation

The Landau-Lifshitz-Gilbert equation (Eq. (11)) is a system of ordinary differential equations (ODEs) that can be written in a general form as *dydt*= *f*(*t*,*y*), *y* – R*N*(38) with the initial condition *y*(*t*0) = *y* 0.(39)

The PVODE package is a general purpose solver for initial-value problems for stiff and non-stiff ODEs of the form of Eq. (38). It is based on CVODE and uses MPI for parallelization and portability. Two methods are available for the numerical solution of Eq. (38): The backward differentiation formula (BDF), which is recommended for stiff problems, and the Adams-Moulton formula for non-stiff problems, both of which feature a variable stepsize and variable order.

At each time step an implicit nonlinear system of equations has to be solved. An efficient method for nonstiff problems is functional iteration, because it does not require the solution of a linear system of equations. However, for stiff problems it is better solved by Newton iteration, which does involve the solu-

tion of a linear system of equations. PVODE uses a Krylov subspace method – the iterative scaled preconditioned generalized minimal residual method (SPGMR), whose performance can be considerably improved with suitable preconditioners. In addition, this method leads to fewer function evaluations of the Landau-Lifshitz-Gilbert equation and allows larger time steps, which gives an excellent performance of the numerical time integration.

The Nudged Elastic Band Method

In order to find possible paths of a micromagnetic system through its energy landscape to a local minimum of the total energy, we have implemented the static energy minimization method (cf. Sec. 3.1) and the time integration of the Landau-Lifshitz-Gilbert equation (cf. Sec. 3.2). However, the investigation of thermal stability, which is an important topic especially in the area of magnetic storage devices, requires the calculation of transition rates between stable equilibrium states of the system. The transition rate between two stable equilibria is determined by the lowest energy barrier (saddle point), which separates them. Henkelman and Jonsson proposed the nudged elastic band method to calculate these minimum energy paths. This method has been successfully applied to complex micromagnetic systems and it is especially suitable for parallelization. The path is represented by a sequence of "images" (magnetization distributions) that connects the two given stable equilibrium states *Mi* and *Mf*.

These equilibria may be obtained using the static energy minimization method, for example. The initial path is given by the initial magnetization distribution *Mi* and the final magnetization distribution *Mf* and a number of images *Mk* in between that can be obtained by simple linear interpolation. Then, an optimization algorithm moves the "elastic band of images" through the energy landscape towards the optimal path, which is defined by (–*E*(*Mk*) · *t*)*t* = –*E*(*Mk*),(41) where –*E*(*Mk*) denotes the gradient of the total energy at image *Mk*and *t* isthe unit tangent vector along the path. In order to avoid kinks in the path the tangent is calculated using forward, backward, or second order central differences. Eq. (41) requires the component of the gradient parallel to the tangent to be equal to the gradient. In other words, the optimal path is characterized by fact that the gradient of the total energy is parallel to the tangent for anyimage *Mk*.

Starting from the initial path an iterative optimization scheme moves the images *Mk* in a direction *D*(*Mk*), which is given by *D*(*Mk*) = –(–*E*(*Mk*) – (–*E*(*Mk*) · *t*)*t*).

The interpretation of this expression is obvious: The images are moved along the negative gradient of the total energy perpendicular to the tangent. The negative gradient determines the direction towards lower energy while the distance between the images is preserved by taking only the component perpendicular to the tangent. Thus, an ordinary differential equation can be formulated *–Mk"t=* *D*(*Mk*),where *t* denotes some artificial time parameter. This ODE is integrated using PVODE. After the discussion of the static energy minimization method and the time integration the Landau-Lifshitz-Gilbert equation, the implementation of the nudged elastic band method has become very simple. The static energy minimization provides us with the gradient of the total energy.

The parallelization of this method can be done by distributing the images. Thus, every processors needs the full set of matrices (but only one copy independent of the number of images), which are required for the calculation of the local fields (and gradients). However, there is no need to partition the finite element mesh any more, because every processor has to do the full calculation for the images. This has the advantage, that no communication is required during the calculation of the gradient of the total energy. Only for the calculation of the tangents the magnetization of some images has to be copied to "neighboring" processors.

Implementation

The implementation of the micromagnetics package is based on the freely available open source toolbox "PETSc" – the Portable, Extensible Toolkit for Scientific Computation.

In addition, PETSc has interfaces to several other software packages. The 'Toolkit for Advanced Optimization" (TAO) is one of them, and it is used for the static energy minimization in the micromagnetics application (in fact TAO itself is built on PETSc). There are also interfaces to ParMetis for parallel graph partitioning and PVODE the parallel ODE integrator. For reasons to be explained later these built-in interfaces have not been used, but still the integration into PETSc is straight forward. The Metis library is used for mesh partitioning, PVODE for the time integration of the LLG equation, zlib for compressed data output, and libpng for the direct output of graphics files in PNG format.

The micromagnetics application can be divided into four parts. First there is a short serial initialization part, which is executed on just one single processor. It reads the configuration files, various input files for the finite element mesh, the initial magnetization, material parameters, etc. If requested, the finite element mesh is refined globally before it is partitioned using the Metis library. Metis uses a very efficient serial multilevel *k*-way partitioning algorithm, which gives high quality partitionings. Each processor then has (approximately) the same number of elements and nodes and a minimum of faces and nodes that are shared with other processors.

The model is cut parallel to the axis of the cylinder, which results in the smallest areas of the cut planes and the smallest number of faces and nodes that are shared by different partitions. To achieve optimal performance, it is sensible to make the initial serial initialization part as short as possible. However, this may require some additional data exchange between different processors at a later stage of the initialization phase. Even worse, it increases the amount of "bookkeeping" that is necessary to keep track of the distributed data. This trade off has shown up in the mesh partitioning phase, which can be done in parallel using ParMetis via the built-in interface of PETSc. Yet, it proved to be easier to handle, if it is done during the serial initialization phase on just a single processor. Since the whole initialization process itself usually takes only a fraction of the time of the rest of the simulation, it is not as important anyway.

According to the partitioning scheme suggested by Metis all data are then distributed to their destination processors. Many parameters like the initial time, number of nodes, elements, and faces, material parameters and vertex coordinates are broadcast to all processors. Other data arrays like the vertex to element assignment, element properties, and other element data are delivered just to those processors that "own" the element. However, these data structures make only a minor part of the total required storage space. All these steps are done only by the first processor, but the serial initialization part is very short and not computation intensive, which is the reason for the excellent scaling of the program initialization.

Then the computation intensive (second) part starts: First, the finite element mesh is analyzed thoroughly. The volumes of all elements and vertices (using the box scheme), element quality factors, and the corresponding extreme and averages are calculated. Next, the stiffness matrix of the Poisson problem and the matrices for the contributions to the local field (Eqs. (21), (28)) are calculated on an element by element basis. However, the results are directly inserted in the global matrices, where the contributions from different elements (even from different processors) are added up by PETSc automatically. This behavior makes the handling of the nodes of the finite element mesh very convenient, because we do not have to manage any duplicate nodes ("ghost points").

The band structure is achieved by a suitable numbering of the nodes, which in this case has been optimized by the mesh generator Patran using the Gibbs-Poole-Stockmeyer algorithm. For the data output along a sampling line through the model and for the graphics output the interpolation matrices are calculated (cf. Sec. 5.2.2). Then, the requested solver (PVODE for integration of the Landau-Lifshitz-Gilbert equation, TAO for energy minimization, or the solver for the nudged elastic band method) is initialized and all additionally required data structures are created.

The initialization phase is finally completed with the calculation of the local fields and energies of the initial magnetization distribution, first entries in the log files, and the removal of data structures that are not needed any more. In the third part the solution of the micromagnetic problem is calculated with the requested solver. All solvers require the calculation of the local effective field Eq. (37) with simple matrix-vector multiplications. Since all matrices (except for the boundary matrix Eq. (36)) are sparse matrices, these matrix-vector multiplications with distributed data are very efficient.

For the integration of the LLG equation or the nudged elastic band method PVODE is called to make one time step. The energy minimization with TAO is performed until equilibrium has been reached. Then all required observables are calculated and written to a log file and additional data of the sampling line and graphics output are stored. Finally, several stopping criteria are checked and the simulation is continued with the next iterative step or it is aborted. When the simulation has been completed, the fourth part of the program is executed. The final results are logged, the solvers are destroyed, all data structures freed and the program exits.

OPTIMIZATION

During the implementation of the micromagnetics application, special attention has been paid to the optimization with regard to speed and storage requirements. The former is strongly determined by good scaling properties of the program on multiple processors. The latter is not an issue for the memory (owing to the distributed data structures), but it can become a problem concerning the size of the data files that are generated and stored on disk during a simulation.

Profiling

The timing of a serial (single processor, single threaded) program is usually measured as the "real CPU time" (the time which is spent executing commands of the program) by the kernel of the operating system. Thus, idle times during which the program is "sleeping" and waiting for other processes or i/o operations to complete are not taken into account. On the contrary, the timing of a parallel program has to be done in terms of the elapsed time between its invocation and termination ("wall clock time"). Idle times of one of the different instances of the program indicate a load balancing problem, which may lead to a less than optimal scaling. As a result, careful profiling requires, that all instances have an equal environment on their processor (i.e. no other processes producing a high load should be running) and all timings have to made in terms of "elapsed wall clock time".

Data I/O

Reading files from and writing them to disk is a quite slow process, but usually the total time spent with i/o operations is negligibly small as compared to the total execution time of the program. More important is the amount of data that have to be stored. Due to the distributed data structures, the size of the finite element models can be increased (almost linearly) with the number of processors, if each of them has the same amount of private memory available. However, the generation of the FE models and the storage of the results to disk can become a problem.

Mesh Refinement

In order to reduce the size of the initial finite element mesh, global or partial (adaptive) mesh refinement can be done at run time. Yet, in parallel programs adaptive refinement methods give rise to another difficulty and that is "load balancing". In order to get the maximum speed up, each processor working for a parallel program should be busy all the time, because delays and idle times reduce the overall performance.

Therefore, the problem is initially split up into parts, which will give each processor the same amount of work. To be more accurate, in a heterogeneous environment where machines and processors with different speeds and capabilities are used, the numerical problem has to be distributed in such a way, that all processors finish their tasks at the same time. Then they do not have to wait for each other, when they have to synchronize their results. Adaptive mesh refinement methods insert and remove finite elements (and therefore nodes) and modify the number of unknowns at run time. Thus, the computational effort will increase for processors working on a partition of the finite element mesh, where elements have been inserted, and decrease for those, where elements have been removed. Of course, this disturbs the initial load balancing and requires a new partitioning of the finite element mesh, redistribution of the current data to the processors and reinitialization before the calculation can be resumed. This scheme has been implemented in the finite element package "UG", whose data model is based on "Dynamic Distributed Data", a software tool for distributed memory parallelization.

Nevertheless, global mesh refinement has been implemented, because it allows much smaller input data files and adds the flexibility of increasing the mesh resolution without remeshing the model with a mesh generator. The global refinement scheme is very simple, because it just splits up every finite element into eight "children" (cf. Fig. 3). If this scheme is applied to the whole finite element mesh, its structure remains consistent. The quality of the mesh is preserved, because the four children in the corners of the parent element are congruent to the parent and the other four from the octahedron in the center have equal volumes, too.

Even though they are not congruent with the parent, it has been shown, that at most three congruence classes occur. Thus, the stability of the generated triangulations is preserved. With every refinement step the number of finite elements increases by a factor of eight and the number of nodes approximately, too. As a result, the input data files and especially the file containing the mesh geometry are much smaller. Even more important, the mesh generator does not have to generate a very fine and therefore large mesh, even though it has to be fine enough to resolve the details of the geometrical model and should give finite elements of good quality. Moreover, convergence of the results can be "easily" checked with a refined finite element mesh.

Data Output

For the direct output of graphics images, the PNG file format has been chosen, ecause it is widely supported (also by all newer WWW clients), generates very small, highly compressed files, and the reference library libpng is freely available.

First, the magnetization, which is defined on the nodes of the unstructured finite element mesh, has to be interpolated on a regular mesh in the desired slice plane. The slice plane is defined in the configuration file in the form $ax + by + cz = d$ with the normal vector $n = (a \mid b \mid c)$. Then all elements that are cut by the slice plane have to be found. In order to simplify this task the whole model (i.e. the Cartesian coordinates of the vertices) is translated and rotated in such a way, that the slice plane coincides with the x-y-plane.

Then all elements that are cut by the x-y-plane – this is easily determined using the z-coordinates of the vertices – are tagged, and the maximum dimensions (the "bounding rectangle") of the finite element model in the slice plane are determined. For a given image size we can now calculate the Cartesian coordinates P of each pixel P, which then has to be assigned to one of the tagged elements, within which the pixel is located. This "point location problem", which is a standard problem in geographic information systems and computer-aided design and engineering, is the most time consuming part.

However, the number of tagged elements (which are cut by the slice plane) is usually far smaller than the total number of elements, and so we can use a brute force yet clever approach that takes advantage of some peculiarities of our specific problem.

In principle, for each pixel we have to search all tagged elements and determine if the pixel is inside or outside. But we can speed up the search by skipping those elements, for which the distance of the pixel from the first node is larger than the maximum edge length of any tetrahedron.

If a suitable element is found we calculate the barycentric coordinates $BP=(\alpha,\beta,\gamma,\delta)$ of the pixel. The barycentric coordinates are defined as the relative position of P inside the tetrahedron $ABCD$. In a "regular" tetrahedron (which has three edges coinciding with the coordinate axes) these correspond to the relative coordinates.

$$P = \alpha A + \beta B + \gamma C + \delta D,\ \delta = 1 - \alpha - \beta - \gamma$$

If we solve for (α,β,γ) we get a simple linear system of equations

$$(\alpha|\beta|\gamma)T = (A - D|B - D|C - D)\text{–}1 \cdot (P - D)$$

The point P is inside the tetrahedron if and only if $\alpha,\beta,\gamma,\delta \geq 0$, which is equivalent to $\alpha,\beta,\gamma \geq 0$ and $\alpha + \beta + \gamma \leq 1$.

The barycentric coordinates $(\alpha,\beta,\gamma,\delta)$ give the weight, by which the data values at the corresponding vertices of the tetrahedron contribute to the linear interpolation at P. Thus, for some data xi defined on the vertices of the tetrahedron, we find the linear interpolation at P with $xP = \alpha x_A + \beta x_B + \gamma x_C + \delta x_D$.

As a result we can assemble an interpolation matrix that calculates the magnetization for all pixels of the image. Let MFE denote the PETSc vector of one Cartesian component of the magnetization of all nodes of the finite element mesh and *M*ip the PETSc vector of interpolated values, then we have *M*ip= *A*ip*M*FE with the ($m \times n$) interpolation matrix ($m = a \times b$ rows for images with $a \times b$ pixels and n columns for a finite element mesh with n nodes). The matrix elements *Aij*,ip are given by the barycentric coordinate of pixel *Pi* corresponding to node with the global index *j* of the finite element, in which the pixel is located. This matrix is very sparse as it has only four entries per row (the four barycentric coordinates of a given pixel).

Since the interpolation matrix depends only on the triangulation, we can calculate it during the initialization phase of the program. Whenever a snapshot of one magnetization component should be stored to disk, a simple matrix-vector multiplication gives the interpolated values at the pixel positions. Then the data are color coded. Finally, the PNG library is called to convert these data into a PNG graphics file.

In addition to snapshots of the magnetization it is often desirable to measure some data along a probing line through the model. This feature has been implemented based on the algorithm described above. For the probing line we translate and rotate the model to make the probing line coincide with the x-axis. Then we tag all elements that are cut by the x-axis and determine the barycentric coordinates of every point on the sampling line. We finally end up with another interpolation matrix that calculates the data values along the probing line. These can then be stored in a data file on disk.

Finally, the complete set of data, which includes the magnetization, the magnetic scalar potential, and the local fields are stored in files in UCD format. This file format can be read by many advanced visualization packages like AVS and MicroAVS. However, the (human-readable) ASCII UCD format generates huge files, because the whole finite element mesh is stored in each file. Thus, two measures have been taken to tackle this problem. First, the finite element mesh is stored in a single separate file in UCD format during the initialization phase. The data files that are generated during the simulation are then stored without the mesh definition, which saves about 50% of disk space.

Secondly, the data files are compressed using the zlib-library in gzip-compatible format, which shrinks the data to about 30% of their initial size. Thus, we end up with data files, which require only 15% of the uncompressed storage space. To restore the data in proper UCD format, a simple shell script uncompresses the data and merges them with the mesh data. Alternatively, there is also a binary UCD format, which gives smaller files than the standard ASCII format, but it is not as widely supported and may lead to compatiblity problems, when the files are transferred to machines with different hardware architecture and operating system.

Performance

The performance (the speedup in particular) of the micromagnetics application has been measured on a Compaq SC45 cluster consisting of 11 nodes AlphaServer ES45 with 4 Alpha processors (EV68 @ 1 GHz, 8 MB Cache/CPU) and 16 GB of shared memory each. The nodes are interconnected with a Quadrics switch, which provides a maximum MPI bandwidth of 600 MB/s. Since this machine has been shared with several other users, up to 24 processors have been available for speedup measurements.

The speedup has been measured as $SP = t1tP$,where $t1$ is the execution time of the program for a given problem on a single processor and tP is the execution time for the same problem on P processors.

The energy minimization method, which uses the LMVM method of the TAO package, has been applied to calculate the nucleation field of FePt nanoparticles.

16 processors we find a “superlinear” behavior of the solution part of the application. This is a well known phenomenon in parallel computing and can be attributed to caching effects. As the same total amount of data is distributed over more processors, the relative amount decreases and may reach a size, where it fits into the fast cache memory of modern computer architectures. As a result, the data need not be fetched from the main memory (which is a lot slower than the cache memory) and the calculations are completed a lot faster. However, as ever more processors are used, communication requires more and more time, which eventually leads to a saturation of the speedup factor. The parallel time integration using PVODE is not as efficiently parallelized as the TAO package.

Permalloy Nanodots

We have studied dynamic precession modes of magnetic vortices in permalloy nanodots, which have also been investigated experimentally by several groups. The hysteresis loops of magnetic nanodots have been measured by vibrating sample magnetometer and magneto-optical methods. Single domain and vortex states have been successfully identified. Furthermore, these magnetic vortex states are an interesting object for high frequency magnetization dynamics experiments, which are important for high-density magnetic recording media, where high-frequency field pulses of the magnetic write head store the information by reversing the magnetization.

We have assumed Js = 1 T for the saturation polarization, $A = 13 \times 10''12$ J/m for the exchange constant, and zero magnetocrystalline anisotropy. For high in-plane external fields, the magnetization is almost uniform. As the field decreases the magnetization distribution turns into an "onion state", which is caused by the magnetostatic stray field. At the nucleation field (about 23 kA/m in our example) the symmetry of the magnetization distribution breaks and a "C" state develops, which leads to the nucleation of a vortex on the circumference. The vortex immediately moves towards its equilibrium position (close to the center of the nanodot). This dynamic process (vortex precession) is discussed below. As a result we find a sudden drop in the average magnetization.

When the external field is reduced to zero the vortex moves into the center of the nanodot. If the external field is increased in the opposite direction, the vortex is forced out of the center of the dot. For about "80 kA/m the vortex is pushed out of the nanodot (annihilation) and we find the second jump in the hysteresis curve to (almost) saturation. A detailed study of the static properties of magnetic vortices will be reported in. This characteristic behavior has also been found experimentally using Hall-micromagnetometry by Hengstmann et al., who measured the stray field of individual permalloy disks using a sub-μm Hall magnetometer.

The hysteresis loops of arrays of Supermalloy nanomagnets have been measured by Cowburn et al. using the Kerr effect. Their characteristic loop shape has then been used to identify the single-domain in-plane and the vortex phase. The dynamic behavior of the magnetic vortex has been studied by applying an instantaneous external field pulse of 80 Oe (8 mT, 6.4 kA/m) in the in-plane direction. The time evolution of)*Mx* (the average of *Mx* over the whole nanodot) for a dot with an aspect ratio of $L/R = 10\ \text{nm}/100\ \text{nm} = 0.1$ is given in Fig. 10(a). Then the damped oscillation, which is caused by the spiral motion of the vortex core towards its equilibrium position, is observed. The corresponding Fourier spectrum is given in Fig. 10(b) and shows a sharp peak at a frequency of 0.7 GHz.

CONCLUSIONS

A scalable parallel finite element micromagnetics package has been developed. It is entirely based on free open source software packages that have been selected for performance, scalability, portability, and ease of use. The combination of static energy minimization, time integration, and the nudged elastic band method makes it a very efficient and versatile tool. Measurements on a parallel computer and a cluster of PCs showed very efficient parallelization and speedup. The program has been used for the investigation of magnetic vortex structures in permalloy nanodots.

3

Spin-polarized Electronic Structure

INTRODUCTION

Nanostructures open new possibilities to tailor the mechanical, chemical, magnetic and electronic properties of materials and, at present, there is strong demand for basic understanding of new phenomena that nano structures may exhibit. Nanomagnetic objects are different from both atoms and bulk materials, thereby providing an interface between physics, chemistry, material sciences, engineering and biology. For example, the length scale of typical nano-structures allows a direct use in many systems, including the human body.

Some phenomena described by first principles are surface-and interface magnetic anisotropies, reduced or enhanced magnetic moments at particle or film surfaces, and optical properties, including magnetic dichroism. In addition, parameters obtained from first-principle calculations serve as an input in numerical micromagnetic simulations.

Here we focus on iron-series transition-metal elements and their alloys. Magnetic oxides and rare-earth compounds can also be used in nanomagnets, but their magnetism goes beyond the scope of this review. For example, the large magnetocrystalline anisotropy of rare-earth transition-metal intermetallics, which are used in nanostructured permanent magnets, largely reflects the Hund's-rules ground state of tripositive rare-earth ions. Magnetic clusters containing a few atoms have been investigated for many decades, both theoretically and experimentally. The electronic structure of atomic clusters was traditionally treated by real-space methods based on various approaches from tight-binding (TB) to density-functional methods.

The methods of calculations of electronic structure of bulk materials use periodic boundary condition and k-space representation. They are very well developed and applied to broad variety of materials. Nano-structures include 100–100000 atoms and do not exhibit periodicity, so that their first-principle study is a very challenging problem. Direct atomistic treatment of real nanodevices is very difficult due to the large number of atoms. The free-electron model, tight binding approaches and order O(N) density functional approach have all been applied to the nanoscale geo metries to study their electronic properties. The choice of an approach depends on the degree of compromise between numerical and conceptual tractability on the one hand and quantitative accuracy on the other.

METHODS OF ELECTRONIC-STRUCTURE CALCULATIONS

Electronic structure methods for studies of nanostructures can be divided broadly into supercell methods and real-space methods. Supercell methods use standard k-space electronic structure techniques separating periodically repeated nanostructures by distances large enough to neglect their inter actions. Direct space methods do not need to use periodic boundary conditions. Various electronic structure methods are developed and applied using both approaches. In this section we will shortly discuss few popular but powerful electronic structure methods: the pseudopotential method, linear muffintin orbital and related methods, and tight-binding methods.

Linear Muffin-Tin Orbital and Related Methods

Basis-set methods may be divided into two categories, depending on whether they (i) use wave function expansions in some set of fixed basis functions like atomic orbitals, Gaussian and plane waves, or (ii) expand the wave function in a set of energy and potential dependent partial waves as done in Korringa-Kohn-Rostoker (KKR) and the augmented plane wave (APW) methods. The KKR and APW methods require computational efforts which, despite attempts to improve the efficiency, are barely feasible and limited to sp-like valence and conduction electrons.

Computationally, this can be remedied by the addition of localized orbitals to the plane wave basis set. But such a hybrid scheme is neither elegant nor in accordance with the chemical and physical intuition based on the smooth trends observed through the periodic table. It is necessary to use the self-consistent methods for computing one electron eigenval-ues and eigenvectors with speed and accuracy. The so-called linear methods of band theory satisfy the requirements rather well. This is true for the linear-muffin-tin-orbital method (LMTO). This method is linearized version of the KKR method. Almost identical with the solid state LMTO method is the augmented spherical wave (ASW) method of Williams *et al.*.

The LMTO has also been extended to treating crystalline impurities with the Green's function technique by Koeing *et al.;* and Gunnarson *et al.* have used it for both metal and semiconductor hosts. Harris, Casula and Herman and Springborg *et al.* have developed the LMTO method for clusters and molecules. For surfaces and thin films, LMTO techniques have been devised by Fuziwara and Femando *et al.*. It was also discovered that the conventional solid-state LMTO basis set can be transformed exactly into orthogonal, tight binding (TB) and minimal basis sets, which simplifies and generalizes the solid-state LMTO method considerably.

The use of the atomic-sphere approximation (ASA), where the single-electron potential is modeled as a superposition of spherical potentials inside overlapping space-filling spheres, makes LMTO methods com putationally very efficient. Where this approximation is applicable, the LMTO-ASA method is presumably the most efficient procedure available for solving the density functional equations to a reasonably high degree of accuracy. However, a full-potential treatment going beyond the ASA is needed for many systems, including surfaces and impurities, and for total-energy changes associated with phenomena such as phonon distortions and atomic relaxations etc.

A number of full potential methods have been developed, namely, FPLAPW, FPLMTO and FPLO. A real-space version of the full-potential approach was implemented by Beck.

Plane-Wave Pseudopotential Methods

The pseudopotential method relies on the separation (in both energy and space) of electrons into core and valence electrons and implies that most physical and chemical properties of materials are determined by valence electrons in the interstitial region. One can therefore combine the full ionic potential with that of the core electrons to give an effective potential (called the pseudopotential), which acts on the valence electrons only. On top of this, one can also remove the rapid oscillations of the valence wavefunctions inside the core region such that the resulting wavefunction and potential are smooth.

Beyond a chosen cutoff radius, the all-electron and pseudofunctions (potential and wavefunction) are identical, while inside the core region both the pseudopotential and pseudowavefunction are smoothly varying. After the construction of these pseudofunctions for a single atom and ensuring that their scattering properties are almost identical to those of the all-electron functions, they can be used in any chemical environment.

The pseudopotential method has various advantages. Eliminating the core electrons from the problem reduces the number of particles that must be considered in the Kohn-Sham (KS) equations for the effective one-electron potential. For example, a pseudopotential calculation for bulk silicon (with 10 core and 4 valence electrons) requires the calculation of 4 occupied bands at each k-point, while an all-electron approach would require the calculation of 14 occupied bands. More importantly, the smooth spatial variation of the pseudopotential and pseudowavefunction allows the use of computationally convenient and unbiased basis, such as plane wave basis sets or grids in space.

The plane-wave pseudopotential method has proven to be an excellent computational tool for solving large scale electronic structure problems. Notable strengths of the method are the ability to use fast Fourier transform to update the Kohn-Sham equations, the lack of dependence on the basis on atom positions, and the clear control of convergence with the cutoff energy determined by the shortest wavelength mode. However, the method encounters difficulties in treating widely varying length scales. This issue is especially relevant for surfaces, clusters, and the hard pseudopotentials of first row elements or transition metals, which vary rapidly near the nucleus.

A real-space version of the pseudopotential method was developed by Chelikowsky.

TIGHT-BINDING METHODS

The traditional approach to obtain the electronic structure of a periodic solid with *ab initio* pseudopotentials has been to solve the Kohn-Sham (KS) equations in momentum space using a plane-wave basis set. There are many advantages of plane-wave basis sets. For example, due to their completeness and orthonormality, they allow convenient and unbiased representation of the charge density and wavefunctions for calculating operator matrix elements.

For non-periodic and localized systems, such as defects, surfaces, clusters, quantum dots, using a plane wave basis set requires an artificial periodicity in order to implement standard electronic structure algorithms. This artificial periodicity is introduced through the use of large unit cells called *supercells*. While the supercell approach works well for localized systems, it is typically necessary to consider a very large supercell. This results in a plane-wave basis replicating not only the relevant electronic states but also vacuum regions imposed by the supercell.

A much more efficient method to implement for investigating the electronic structures of localized systems is to use real space methods such as the recursion methods and the moments methods. These methods do not require symmetry and their cost grows linearly with the number of inequivalent atoms being considered. For these reasons, real space methods are very useful for a description of the electronic properties of complex systems, for which the usual k-space methods are either inapplicable or extremely costly.

Real-space methods are generally impractical, but they are extremely efficient when the system in consideration is well described by a tight-binding (TB) Hamiltonian. For a long time, parameterized TB Hamiltonians in conjunction with real space methods have been used to study the properties of complex transition metal alloys. In some cases, simple model TB Hamiltonians are very useful to stress the essential physics of the phe nomenon in question and even to give the correct trends found in real systems.

But in other cases, the effects are too subtle and the results obtained by the use of these parameterized Hamiltonians can be misleading. Even when trends can be well described, to obtain quantitative results a more exact Hamiltonians is usually needed. This has been a major problem regarding real-space electronic structure calculations. Usually, the involved parameters needed to build the TB Hamiltonian, such as hopping integrals, are obtained from a linear combination of atomic orbitals (LCAO) fit to more exact k-space calculations or are adjusted to fit experimental results.

When one considers complex systems, for which no exact results are available, parameters are usually obtained from a simple similar system and then transferred to the more complex system in question using the Slater-Koster rules and some other empirical formulas which are known to be roughly obeyed. The results were often encouraging, but the lack of a solid theoretical background to justify the procedure left some fundamental questions unanswered. For example, how could one simulate the crystal-field splitting which depends on local environment and can not be transferred from one calculation to another, and how do we treat the wave function?

The solution to these problems is not unique and several interesting points of view have developed. However, major progress towards obtaining a TB-Hamiltonian based on a solid theoretical understanding of the problem came in 1984 with the advent of the LMTO-ASA-TB formalism. This gave rise to a new era regarding the quantitative description of the electronic structure of complex systems by using real-space methods.

MAGNETIC PROPERTIES

In this chapter we describe how basic or "intrinsic" magnetic properties are calculated from first principles. Technological applications, for example in magnetic recording, tend to require a large magnetization, thermal stability, and control over the coercivity. The reversal process is described usually by micromagnetic modeling which takes into account dipole-dipole inter actions. The description of magnetization reversal is beyond the scope of our paper.

Magnetic properties can be understood and fully described from microscopic considerations using *ab-initio* electronic-structure methods. The local magnetic moment in combination with intra-atomic exchange defines the magnetization, interatomic exchange describes the thermal stability, and the coercivity reflects the magnetic anisotropy of the system. These interactions have different energy scales. The intra-atomic exchange is of the order of 2 eV, interatomic exchange of the order of 10 meV, and anisotropy of the order of 1 meV.

However, all of them are extremely important to describe the behavior of a magnet. The description of nanomagnets requires new approaches. First, nanostructures are not periodic and tend to have large surface-to-volume ratios. Because of this the magnetization is not uniform across the nanostructure, local magnetic moments differ from site to site, exchange coupling varies through-out the nanostructure, and the anisotropy can be quite different from bulk or surface anisotropies. Second, it is hard to define properties in the similar fashion as in the bulk or as in case of molecules.

Magnetic Moment

The magnetic moment is given by the difference between the filling of majority and minority spin bands, which is directly related to the ferromag netic exchange splitting between the bands. There is also an orbital-moment contribution to the moment, but in 3d-based magnets, it is often small due to quenching. Total energy minimization over all occupied band states yields crystal structure and magnetic ordering. Magnetic coupling and magnetic moment are sensitive functions of the interatomic spacing and the atomic volume, respectively. Both are dominated by the Coulomb and by exchange interactions at small distances. The moment is determined pri marily by intraatomic exchange interactions and the Curie temperature by the interatomic exchange interactions.

The most common approach to calculate the magnetization from first principles is to use the local density formalism, where the spin dependent part of the total energy is approximated by an expression derived for the electron gas. The corresponding exchange-correlation potential depends only on the majority and minority spin densities. Different magnetic struc tures, such as paramagnetic, ferromagnetic and various anti-ferromagnetic configurations, exhibit different total energies, and the lowest energy configuration obtained by local density theory represents the observed magnetic state in most cases. Room-temperature ferromagnetism occurs in rather few elements, Fe, Co, Ni, Gd, and several other rare earths are ferromagnetic at low temperatures. However, many alloys are ferromagnetic.

The stability of the ferromagnetism of itinerant magnets can be explained by the *Stoner criterion,* which takes the DOS at the Fermi level, $D(EF)$, and an atomic exchange integral I as input. The transition from paramagnetism to ferromagnetism is favourable for $D(EF)$ $I > 1$. Then the system can lower its energy by bringing enough majority spin electrons down in energy by opening up the ferromagnetic exchange splitting.

The Stoner criterion explains why Fe, Co and Ni are singled out for ferromagnetism. Several other elements are close to fulfilling the criterion, for example Pd. Since the DOS is different for different atomic sites, nanostructures may exhibit different local moments and complicated magnetic transitions.

With increasing atomic volume, one approaches the free atom limit where Hund's first rule postulates maximum spin, so that the individual spins of the electrons in a shell are aligned parallel.

More generally, Pauli's exclusion principle implies that electrons with parallel spins have different spatial wavefunctions, reduces the Coulomb repulsion and is seen as exchange interaction. When the atoms are squeezed into a solid, some of the electrons are forced into common spatial wavefunctions, with antiparallel spins and reduction of the overall magnetic moment. At surfaces and interfaces, the reduced coordination reverses this effect, and a part of the atomic moment is recovered.

Anisotropy

Magnetic anisotropy is one of the most important properties of a magnetic material. It is the energy necessary to deflect the magnetic moment in a single crystal from the easy to a hard direction. The main contribution to the anisotropy is *magnetocrystalline anisotropy.* The easy and hard directions arise from the interaction of the spin magnetic moment with the crystal lattice that is due to spin-orbit coupling.

In crystals, the magnetocrystalline anisotropy energy is given by a series expansion in terms of the angles between the direction of magnetization and the crystal axes. It is often sufficient to represent the anisotropy energy in an arbitrary direction by just the first two terms in the series expansion. These two terms each have an empirical constant associated with them called the first- and second order uniaxial anisotropy constants K1 and K2 respectively. In cubic crystals, the lowest-order term is the fourth-order term, also known as the "cubic" K1.

There is also some shape anisotropy due to magnetostatic dipole interactions. This anisotropy is important in some nanomagnets, for example in elongated nanoparticles, but unrelated to the electronic structure. A contribution caused by spin-orbit coupling and closely related to magnetocrystalline anisotropy is *magnetoelastic anisotropy.* Mechanical stress creates a strain which amounts to a lattice distortion and yields a correction to the magnetocrystalline anisotropy. Surface anisotropy is a manifestation of magnetocrystalline anisotropy, too.

Starting with Brooks (1940), calculations of the magnetocrystalline an-isotropy energy (MAE) of itinerant magnets have developed in two different approaches have been developed since the pioneering studies of Takayama,

Bohnen, and Fulde and of Gay and Richter. The first approach uses perturbation theory within a semi-empirical tight-binding framework, justified by the argument that the MAE is small compared to a characteristic bandwidth. The second approach relies on *ab initio* spin-polarized total-energy calculations, with spin-orbit coupling included either self-consistently within the scalar-relativistic approximation or as a final perturbation to a calculation neglecting spin-orbit coupling and using the force theorem.

The main difficulty with this approach arises from the fact that a change in the direction of the magnetic moment changes the occupation of the eigenstates only in a narrow region of the Brillouin zone, hence the k-space integrations for the total energy are only very slowly convergent. To avoid the necessity to use a huge number of k-points, Wang, Wu, and Freeman introduced a state tracking procedure, using information on the change of the band structure with increasing spin-orbit interaction to extrapolate the Brillouin-zone integrals.

The description of the magnetic anisotropy of nanoparticles is more complicated than that of bulk systems, partially because shape and surface anisotropies are of the same order of magnitude as the magnetocrystalline anisotropy. The ratio of atoms on the surface to the number of atoms in the core of a nanoparticle is very large.

This changes the electronic structure of the core atoms as well and its anisotropy may be quite different from the bulk systems. Surface anisotropy may cause drastic change in magnetization and reverse properties of nanostructures. Even in a naive magnetostatic approach, surface anisotropy causes the renormalization of switching parameters.

SPIN STRUCTURE AND NONCOLLINEARITY

The term spin structure refers to the ferromagnetic, antiferromagnetic or ferri-magnetic alignment of the atomic spins. In ferromagnets, all spins are parallel and the atomic moments add, whereas ferromagnets and antiferromagnets are characterized by two or more sublattices with opposite moments. This leads to a reduction or absence of a net magnetic moment. Competing exchange interactions in periodic crystals and in disordered magnets give rise to noncol-linear spin arrangements. Examples are helimagnetic order in perfect crystals, which is caused by the competing interactions between next and more distant neighbors, and spin-glass behaviour in magnets with atomic scale disorder.

Deviations from parallel or antiparallel spin alignment may also occur at surfaces and interfaces. Further, it is important to distinguish between zero temperature magnetism and finite temperature magnetic order. It is difficult to calculate finite-temperature magnetic properties from first principles, particularly Curie and Neel temperatures. In the atomic limit, for example using the Heisenberg model, one assumes a magnetic moment on each atom that persists beyond Curie temperature. In the paramagnetic phase, the magnetization vanishes by orienting the moments at random.

The other extreme, that is the Stoner model of band like ferromagnetism, assumes that the magnetic moment gradually decreases at elevated tem perature and vanishes at the Curie temperature. It yields Curie temperatures that are several times higher than observed. The basic drawback is the assumption that the magnetic order disappears by spin flip excitations across the Stoner gap. They cost more energy than is actually needed to disorder magnetic moments.

More realistic models contain small domains with local order whose orientations are starting to fluctuate when approaching the Curie temperature. Some degree of local correlation remains even above TC. Such a picture brings in spin wave excitations (magnons) which are not contained in the ground state energy bands. The efforts to model the spin structure just below and above the Curie point from first principles give qualitatively correct Curie temperatures, even if the method used in the paper is more appropriate to very weak itinerant ferromagnets such as $ZrZn^2$.

Interatomic Exchange

Magnetic order occurs because the magnetic moments of neighbouring atoms couple by interatomic exchange. As the intra-atomic exchange responsible for the magnetic moment, it reflects the overlap of electrons in combination with Pauli's exclusion principle. Depending on the positive or negative sign of the exchange, the interaction between the ions will force the individual moments into parallel (ferromagnetic) or antiparallel (antiferromagnetic) alignment.

The direct exchange between overlapping orbitals was proposed first. Later other types of interatomic exchange were identified such as an indirect or RKKY-type exchange, superexchange, double exchange, anisotropic exchange, and higher-order exchange (such as biquadratic).

All of them may be present in a system and responsible for a variety of effects governing the magnetic structure and finite-temperature behaviour of magnet. Direct exchange operates between moments which are close enough to have sufficient overlap of their wavefunctions. it gives a strong but short range coupling which decreases rapidly as the ions are separated.

Direct exchange as originally envisaged by Heisenberg, is always ferromagnetic, but there is a negative hopping contribution. When the atoms are very close together, the electrons minimize the single-electron interaction by spending a considerable time between the nuclei. Since the electrons are then at the same place in space at the same time, Pauli's exclusion principle requires them to possess opposite spins. This gives rise to antiparallel alignment, corresponding and therefore negative or antiferromagnetic exchange. Note that the direct intra-atomic exchange between electrons in the same atom J is positive, leading to Hund's first rule. Indirect exchange couples moments over relatively large distances. It is the dominant exchange interaction in metals where there is little or no direct overlap between neighbouring magnetic electrons. It therefore acts through an intermediary which in metals is the conduction electrons (itinerant electrons). This type of exchange is also known as the Ruderman, Kittel, Kasuya and Yoshida (RKKY) interaction, especially when mediated by free electrons.

This spin polarization in the itinerant electrons is felt by the moments of neighboring magnetic ions within range, leading to an indirect coupling. Since RKKYexchange oscillates from positive to negative as the separation of the ion changes, the magnetic coupling can be ferromagnetic or antiferromagnetic. In rare-earth metals whose magnetic electrons in the 4f shell are shielded by the 5s and 5p electrons, direct exchange is rather weak and insignificant and indirect exchange via the conduction electrons gives rise to magnetic order in these materials.

Electronic Structure of Nanomagnets

Recent advantages in the preparation and characterization of thin films, surfaces and nanostructures have lead to an intriguing question: is it possible to fabricate 'designer solids' by controlling materials on the atomic scale, that is layer by layer, row by row, and ultimately atom by atom? Engineered molecules are common place in biochemistry, and the same idea can be apply to solids and electronic materials. Electronic properties of semi conductor devices have been controlled by heterostructures, quantum wells and superlattices.

Magnetism as a cooperative phenomenon also lends itself to manipulation in small structures, where neighboring atoms can be replaced systematically by species with stronger or weaker magnetism. In fact, a class of magnetic/non-magnetic multilayers termed "spin valves" has been introduced into magnetic storage devices. The wave functions of electrons undergo changes on confining them to dimensions comparable with their wavelength. In semiconductors, confinement and quantization phenomena are visible already at dimensions greater than 200 nm, whereas in metals they typically are seen at 1 nm. In fact, the Fermi wavelength of typical metals has atomic dimensions, but beat frequencies with the lattice can be an order of magnitude larger.

Consider the formation of low dimensional electronic states by quantization. Confining electrons to small structures causes the continuous bulk bands to split up into discrete levels, for example quantum well states in a slab. For N atomic layers in the slab there are N levels. In order to exhibit two dimensional behavior there should be only a single level within about $6k$B T of the Fermi level.

Several levels within the Fermi cutoff would already approach a three dimensional continuum. For a coarse estimate of the corresponding slab thickness, one may set the energy E of the lowest level equal to kB T. For room temperature ($E = k$B T = 0.026 eV), one obtains a de Broglie wavelength $h/p = h/(2mE)$ ½ 8 nm, which is comparable with the spatial extent of the lowest quantum state. Thus, both the high electron density in magnetic metals and the requirement of room temperature operation for quantum devices point to dimensions of a few nanometers. Note, however, that Fermions are quantum-mechanical objects at any temperature below EF$/k$B.

The contribution of first-principle calculations to physics, chemistry, materials science, and, recently, geology and biology is more important than ever. The main reasons are the steady increase in computer power and the continuous progress in methodology (both efficiency and accuracy of algorithms and approximations). However, the application of these methods to investigate structural, electronic, and optical properties of nanostructures has not been straightforward due to extensive computational needs and to the new physics inherent in the nanometer and subnanometer region.

One particularly useful extension to overcome the computational demand imposed by localized nanostructures has been the introduction of methods based on a real-space implementation, such as the higher-order finite-difference pseudopotential methods. In the most favorable cases, quantum mechanical formulations of the electronic structure of atomic systems scale as the cube of the number of atoms of the system. This makes it very difficult to reach system sizes larger than a few hundreds of atoms, and is therefore a huge barrier for the study of problems in nanoscale materials.

The way out of this trap was found several years ago, when a number of ideas suggested the possibility of developing approximate although accurate schemes to reduce the computational cost to linear scaling. These so called $O(N)$ methods have matured since those first proposals, and now constitute a viable route for studying systems with unprecedented size.

FIRST-PRINCIPLE STUDIES OF NANOSTRUCTURES

Recent advances in theoretical methods and high-performance computing allow for reliable first principle calculations of complex nanostructures. Nanostructured materials are characterized by a fascinating diversity of geometries, but here we restrict ourselves mainly to first-principle cal culations for nanoparticles and clusters, nanowires and nanocontacts. Nanoscale multi-layers are also discussed very briefly, although multilayers are often considered as a subfield of thin-film physics rather than nanoscience. We also ignore nanotubes, because most of the work in this direction has been done on non-magnetic carbon nanotubes.

Nanoparticles and Clusters

In magnetic recording, a bit of information corresponds to a small area or volume of ferromagnetic material with a well-defined and as high as possible magnetic moment. Furthermore, below a given size the grains tend to become superparamagnetic at room temperature, and the stored information would become unstable. It is known, both experimentally and theoretically, that the magnetic moments and the magnetic anisotropy energy per atom are often enhanced in small metallic clusters, as compared to the bulk crystalline state. Nanosize transition-metal particles, between the molecular and metallic states, are of particular interest in this context due to their intrinsically high magnetic moment. The properties of these clusters depend strongly on its surrounding. We will compare magnetic properties of free-standing clusters,

clusters supported on substrate, and clusters embedded in the media.

Studies of small free-standing FeN, CoN and NiN clusters have revealed that their magnetic moments are significantly larger than the corresponding bulk magnetizations. Non-vanishing magnetic moments have even been observed in small clusters of some 4d transition metals, for example Pd, Ru and Rh, which are non-magnetic in the solid state. For example, Kumar *et al.* performed ab initio calculations on the growth and magnetic behavior of Pd clusters having up to 147 atoms and found an icosahedral growth and a ferromagnetic behavior in these clusters.

The magnetization of free clusters of elemental metals exhibits an oscillatory behavior with increasing cluster size. The origin of these oscillations is the successive addition of atomic shells with increasing cluster size. The variation of the magnetic moment in the outermost shell is due to the changing coordination number of surface atoms. As a rule, surface atoms with lower coordination (lower number of nearest neighbors) have larger moments, and the corresponding theory of magic numbers seems to explain experimental observations.

Using a tight-binding Hubbard Hamiltonian in the unrestricted Hartree-Fock approximation, Pastor *et al.* have shown the size and structural dependence of magnetic properties of free Crn, Fen, and Nin ($n < 15$) clusters, and also the exchange interaction and local environments effects in Fen clusters. The finite-temperature magnetism of small clusters, which is remarkably different from that of bulk systems, has also been studied in terms of a similar approach by taking into account both electronic and structural excitations. A great advantage of the TB methods is that they easily can be combined with molecular dynamics calculations, enabling the investigation of relaxation effects which proved to be important in determining the magnetic moments and the MAE of transition metal clusters. As mentioned above, the anisotropy of the free standing clusters is very complicated. Experimental studies predict a very complicated switching behavior, suggesting a strong competition between different contributions to the anisotropy energy.

The main reason is that different pairs of spin-orbit split levels yield anisotropy contributions of opposite sign, so that the net anisotropy is the sum of many positive and negative contributions. Phenomenological studies complement this point and yield good agreement with experiment. The supercell-based studies of supported clusters show that the magnetic moment tends to increase as well. For example, small Co nanoclusters may exhibit a large orbital moment and large anisotropy on Pt surface.

The pros and cons of large orbital moments in applications of nanostructures was discussed by Andriotis *et al.*. Magnetic-nanoparticle patterned media are extensively studied as a potential breakthrough technology for high-density storage of information. The key requirement is that the individual nanoparticles exhibit both a large magnetic moment and a high anisotropy. More generally, in the last decades, theoretical and experimental studies have shown that the reduction of size and dimensionality can give rise to a large range of novel materials. This has opened the possibility of generating specifically designed nanomaterials with tailored properties. Understanding the conditions for the development of such properties is crucial in order to be able to properly adjust and control these parameters, in particular for magnetic applications.

The embedding technique based on the Korringa-Kohn-Rostoker (KKR) Green-function method in the local spin-density approximation (LSDA) has been applied to the magnetism of transition metal adatoms and clusters deposited on surfaces. The main feature of this approach is that the interaction between adatoms and host surface atoms can be analyzed within first-principles electronic structure calculations.

An accurate cal culation of the total energy in terms of full potential or full charge density schemes made possible the investigations of the energetics of adatoms. As compared to TB methods, an obvious drawback of the embedded KKR technique is that, with respect to computational limitations, the number of the atoms in the cluster is restricted to about less than 100. Furthermore, the inclusion of structural relaxations is exceedingly difficult. On the level of a fully relativistic spin-polarized electron theory, recently, strongly enhanced orbital magnetism and MAE of adatoms and small clusters on Ag and Au (100) surfaces have been reported.

Spin polarized electronic structure calculations of the Co clusters embedded in Cu matrix have been reported by Qiang *et al.*. The authors found that the magnetization of the embedded Co clusters in Cu matrix is somewhat lower than the magnetization of the bulk Co, *i.e.* the effect is opposite to the free cluster case. This is caused by the slight suppression of the moment of cobalt sites in the interface region. The magnetization oscillations are similar to the free case in theory if shell by shell filling of the clusters is expected. However, this condition is not satisfied in case of embedded clusters and magic number theory does not work.

Moreover, there is always some interdiffusion on the interface between atoms of the matrix and atoms of the cluster in case of larger clusters. Frequently it creates a magnetically dead layer at the interface. The exchange interactions between metallic clusters in metallic media exhibit oscillatory dependence on the intercluster distance like it happens in Co clusters embedded in the Cu matrix. The calculated inter-cluster exchange interactions shows both ferromagnetic and aniferromagnetic exchange between clusters, and the values of exchange at short distances is of the order of interatomic exchange in the bulk cobalt. The spin-glass type behavior in this and similar systems arises from frustration.

This is very likely responsible for the lowering of the magnetization with the increase of cobalt concentration from 10 to 50%. Nanostructured permanent magnets use the idea of mixing soft magnetic materials and hard magnetic materials in a composite structure. The goal is to combine a large magnetocrystalline anisotropy of hard magnetic materials and large magnetization of the soft magnetic materials in order to obtain large energy product (*BH*)max, which requires large magnetization and large coercivity. This idea was first proposed by Kneller and Hawig, by estimating the effect in multilayers. Micromagnetic calculations based on parameters obtained from first principles show that FePt/Fe nanostructures may exhibit a (*BH*)max as high as 90 MGOe. These changes are experimentally observed in various systems, and Section 2.3 in the following chapter discusses some systems.

The properties of embedded nanoclusters can be controlled by the properties of the matrix. For example, the matrix can exert uniaxial or biaxial strain on the embedded nanoclusters. The corresponding strain can modify the properties of nanostructure considerably. For example, Zeng *et al.* show that applied stress can modify the Curie temperature and anisotropy of FePt.

Nanowires

The progress in experimental techniques has made it possible to synthesize stable ultrathin metal nanowires with diameter down to atomic sizes and of high aspect ratios. An example is single-crystalline silver nano wires having a width of 0.4 nm width and deposited into the pores of an organic template. Kondo *et al.* have observed novel helical multishell structures in the suspended ultrathin gold and platinum nanowires. This is a relatively mature field of studies with thousands of articles published in the last few years, and there are various review articles available.

While magnetic properties of free-standing nanoclusters have been extensively studied experimentally, studies of magnetic nanowires are largely limited to theoretical calculations. One exception are wires freed from templates, where mechanical stability excludes very thin diameters. Decades ago, Weinert and Freeman investigated the electronic structures and magnetism of the linear Fe and Ni chains by using the self-consistent full-potential linearized augmented plane wave (FLAPW) method. They predicted large magnetic moments of 3.3 ìB and 1.1 ìB for the Fe and the Ni chains, respectively. Free-standing Coatomic chains show an en hanced magnetization of 2.3 ìB as well.

It is interesting that first-principle calculations predict ferromagnetism in monatomic nanowires of the nonmagnetic transition metals Ru, Rh, and Pd, with respective mean-field moments of 1.1, 0.3, and 0.7 ìB at the equilibrium bond length. An analysis of the band structures indicates that the nanocontact superparamagnetic state suggested by calculations should affect the ballistic conductance between tips made of Ru, Rh or Pd, leading to possible temperature and magnetic-field dependent conductance. One method to produce nanowires is the growth on the steps of the crystalline surface cut with the high Miller index.

Spisak and Hafner investigated the position dependent magnetic moments of the Fe nanowire on a vicinal Cu(117) by using the pseudopotential method. From the total energy comparison, they found that the system of the Fe nanowire placed on the inner comer site of the Cu(117) step is most stable with a magnetic moment of 2.80 ìB /atom, while the system of the Fe nanowire placed on the top of the terrace (formed by (001) plane) of Cu(l 17) is unsta ble with a magnetic moment of 2.96 ìB/atom.

Jin *et al*. investigated the electronic structures and magnetism of an Fe nanowire along the direction on a Cu(00l) surface (Fe chain) by using the FLAPW method and found large magnetic moments of the Fe atoms:

3.11 and 3.39 ìB /atom for the Fe [010] chain and the free-standing nanowires (free-Fe[010] chain), respectively. Since free-standing nanowires of elemental metals are one-dimensional objects, they exhibit magnetic properties similar to free standing clusters, *i.e.* its magnetic moment increases especially for the atoms at surface. The lower coordinated atoms have higher moment, while atoms inside of the nanostructure have magnetic moment close to the bulk values.

Recently, investigations of the electronic structures and magnetism of an Fe chain along the direction on a Cu(001) were performed by using the FLAPW method within the generalized gradient approximation. Hong *et al.* have calculated properties of Co atomic wires on the Cu(001) surface. They show that magnetic moment of atomic size wires can be affected by the substrate. The magnetocrystalline anisotropy of Co wires supported on Cu substrate has opposite sign from the free standing case, i.e. the magnetization is along the wire on the Cu surface and perpendicular in free-standing case. Gambardella *et al.* investigated a quasi-one-dimensional Co chain at the

Pt(111) step edge. The symmetry breaking at the step leads to an easy magnetization axis at an odd angle of about 20 degrees towards the Pt step. The orbital and spin moment may have different orientation in this system. This reflects the specific character of the crystal-field interaction in these systems.

It is difficult to tackle the problem of embedded nanowires. An example is Co nanowires fully coated with atomic Cu shells. Wang *et al.* find that Cu atoms occupy the surface, while Co atoms prefer to occupy the interior of the nanowires to form the perfect-coated structures. The coated Cu atomic layer leads to a large variation of the magnetic moment of the Co nanowires, depending on the structure and the thickness of Cu layers.

A single row of Co atoms in the center of nanowire is found to be nonmagnetic when coated with two Cu layers, while all other investigated Co nanowires remain magnetic, albeit with reduced magnetic moments. The transport properties of nanowires are of technological importance and have attracted significant attention in the recent years. Band structure gives simple solution to the analysis of the ballistic transport of periodic nanowires because the number of the bands crossing the Fermi surface is equal the number of quantum of conductance. However, the situation in nanocontacts is more complicated.

Nanocontacts

The understanding of spin transport in order to use it in electronics devices has been a field of intense research for the past few years. Interesting effects were reported in electrodeposited Ni nanocontacts, where resistance changes by a huge value on application of a small magnetic field at room temperature. This has followed after effort was focused on the magnetoresistance effect of electrodeposited nanocontacts. Smaller magnetoresistance ratios, but interesting from the point of view of applications, were reported also for Co, Fe and Fe_3O_4 nanocontacts.

There have been several and partially exclusive attempts to explain the experimental data. One possibility considered is the scattering of the electrons in a domain wall. The width of a domain wall in a nanocontact is predicted to be of the order of the nanocontact size and such a reduction of the domain wall width is likely to enhance the magnetoresistance. However, this effect is very small, unless the domain walls contain atomically sharp feature, especially layers with reduced exchange. Magnetostriction effects could also lead to an increased magnetoresistance, but most probably cannot explain the experiments.

Alternatively, a thin domain wall pinned in a magnetic dead layer together with the possibility of 100% polarization of the Ni d electrons was also proposed as a possible explanation. Using a free-electron model and a linear response theory Zhuravlev *et al.* investigated a spin-dependent electronic transport in magnetic nanocontacts in the ballistic regime of conduction. In particular, they show that the presence of a nonmagnetic region in the nanocontact separating two ferromagnetic electrodes can lead to a spin blockade resulting in very large values of magnetoresistance. Other possible explanations are a large magnetostriction in the experimental samples, or the presence of the defects in the constriction. The latter may result in the resonant tunneling and provide large MR.

In order to describe the nanocontacts theoretically, several methods have been developed by various groups. Recently, Burton *et al.* investigated the electrical resistance of a constrained DW in a nanocontact using micromagnetic modeling and ballistic conductance calculations. They found that for electrodes with significantly different magnetic anisotropy, the DW can be trapped by the nanoconstriction (domain-wall pinning). They showed that the DW width and, therefore, the conductance can be controlled by an applied magnetic field. The first-principle approach is based on the tight-binding (TB) formulation. An important conclusion coming from TB models is that conductance of the single-atomic contacts is proportional to the number of valence orbitals available at the Fermi energy. *Ab-initio* density functional (DF) calculations with jellium electrodes have been proposed by Lang *et al.* and Kobayashi *et al.,* which addressed single-atom contacts, atomic chains of Al, Na, and C.

The formation mechanisms of atomic chains made from different types of elements such as Ni, Pd, Pt, Cu, Ag and Au were studied by means of molecular dynamic simulations. Also, Mehrez *et al.* and Brandbyge *et al.* presented fully self-consistent DF calculations of the conductance of atomic contacts treating the electronic structure of the whole system (electrodes and the constriction region) on the same footing. The ballistic conductance through Ni, Co, and Fe nanocontacts within a semiempirical tight-binding model proposed by Velev and Butler shows that the interplay of the contact and the domain-wall resistance can produce very large giant magnetoresistance ratios.

However, the very giant magnetoresistance is limited to geometries in which the nanocontacts are very narrow and have very small aspect ratios, similar to the predictions in. Solanki *et al.* calculated the electronic structure and conduc tance of atomic-size Ni contacts using a real-space tight-binding LMTO method and recursion technique within the frame work of density functional theory. They used the Landauer-Buttiker approach to calculate the conductance. Their results of spin-dependent conductance as a function of energy show ballistic bulk-like behavior. The appreciable dependence of the conductance on the structure of nanocontact is reported in their work. Similar conclusion was made by Emberly *et al.* considering carbon nanowires. We should notice that there are several research groups which could not reproduce giant MR in the nanocontacts experimentally.

Multilayers, Surface, and Interfaces

Magnetic thin films and multilayers have attracted much interest in the context of nanomagnetism, although they are often considered as a separate branch of condensed matter physics. Ultrathin magnetic films, epitaxially grown on nonmagnetic substrates, are prototype systems for inves tigating magnetism in two dimensions, and the possibility to produce films and multilayers with perpendicular magnetic anisotropy for magnetooptic recording applications is of technological importance. The topic of perpendicular magnetic anisotropy is very challenging, since the magnetostatic dipolar interactions always prefer in-plane orientation of the easy axis of magnetization.

Since the pioneering experimental work by Gradmann and the theoretical predictions of Gay and Richter much effort has been spent investigating Fe, Co, and Ni films grown on noble-metals substrates. Among the hetero-epitaxial systems that show strong per pendicular magnetic anisotropy (PMA) are Co/Au(111) thin films and superlattices, Ni/Cu(001) and Ni/Cu(111) thin films, Co/Au(111) thin films and multilayers, bcc Fe/Ag(001) and fcc Fe/Cu(001) thin films. Sabirianov and Jaswal carried out first-principles studies of exchange-coupled hard-soft multilayers predicting very large energy products needed for hard-magnet applications.

The discovery of the giant-magnetoresistance (GMR) effect has further renewed interest in the theory of transport in layered and granular systems. Several approaches to transport in layered systems are currently being used. The phenomenon of GMR is promising for its applications, such as in nonvolatile magnetic random-access memories in the information technology industry or as reading heads and various kinds of sensors in the recording and car industries, respectively. The GMR effect has been observed mostly in the diffusive transport regime in which the mean free path is much smaller than the dimension of the so-called active part of the multilayer system. In the ballistic regime, the mean free path is larger than the dimension of the active part of the multilayer system. Transport in multilayers has been studied extensively and review on the subject covers the topic thoroughly.

The magnetocrystalline anisotropy of multilayered systems has been also calculated by many groups. Újfalussy *et al.* performed fully relativistic spin-polarized local spin density calculations for Co monolayer on Au(111). They obtained an enhancement of perpendicular magnetic anisotropy as a function of the Au coverage. They showed the close relationship between the anisotropies of orbital magnetic moments and the anisotropy energies, thus interpreting their results in terms of familiar perturbation theory.

Lorentz and Hafner presented calculations of the magnetic structure and of the uniaxial and planar anisotropies of thin films of Fe on Cu(001) substrates. They have used the real space recursion calculations using TB-LMTO to calculate the anisotropy. Their results for Fe monolayers reveal the importance of the surface and interface anisotropies, including the effect of nonmagnetic coverages. Essentially, this is a crystal-field effect, similar to that in bulk compounds consisting of atomic layers with different magnetic properties.

Stepanyuk and Hergert have performed molecular dynamics simulations of Co nanostructures on Cu(001) and Cu(111) surfaces. They first performed *ab-initio* calculations of the electronic structure in the framework of density functional theory and then used the results for dynamical simulations.

Summary

One of the goals of computational nanoscience is to calculate physical and chemical properties from first principles. This requires the knowledge of the electronic structures of the materials system in question. The density-functional theory (DFT) makes a huge step towards this goal, by reducing a highly complex many-electron problem to that of noninteracting particles under the influence of an effective potential. The adiabatic approximation allows one to separate the ionic degrees of freedom from those of the electrons. However, in order to apply DFT in practice one has to resort to approximations for electron exchange and correlation such as the local-density approximation (LDA) or the generalized-gradient approximation (GGA). In this chapter, we have outlined some basic principles and some important examples of this approach, representing a very large and growing body of experimental and theoretical work on magnetic nanostructures such as clusters, nanowires, nanocontacts, and multilayers.

4

Nanomagnetic Models

INTRODUCTION

Nanomagnetism involves both atomic and macroscopic features, but it cannot be reduced to a superposition of these two limits. Macroscopic interactions, as exemplified by the compass needle and the geomagnetic field, and atomicscale magnetic phenomena, such as quantum-mechanical exchange, are necessary but not sufficient to understand nanomagnetism. Solid-state matter is made from atoms, and the corresponding length scale, Bohr's radius $a_0 = 0.52$ Å, determines the range of the exchange responsible for bonding and magnetism. On the other hand, Maxwell's equations are macroscopic, that is, they do not correspond to any characteristic length. Why are the phenomena considered in this book realized on a length scale of a few nanometers, as contrasted to, for example, a few meters?

An answer is provided by the relativistic nature of many magnetic phenomena of importance in nanomagnetism. For example, typical domain walls have a thickness of few nanometers. This length reflects the competition between nonrelativistic exchange and relativistic magnetic anisotropy. A simple but qualitatively correct picture is obtained from the relativistic electron energy $mc^2 \sqrt{1 + v^2/c^2}$, where v is the electron velocity. Expanding the energy into powers of v/c yields the rest energy $mc2$, the electrostatic or 'nonrelativistic' energy $mv2/2$, and the lowest-order elativistic correction $(\alpha/2)2mv2/2$, where $\alpha = 4\pi\varepsilon o e2/hc$ is Sommerfeld's fine-structure constant. Here we have exploited that typical electron velocities in solids are of order $v = \alpha c$. Respective examples of nonrelativistic and relativistic magnetic interactions are exchange, which has the character of an integral over electrostatic interactions, and spin-orbit coupling, which leads to magnetocrystalline anisotropy.

On an atomic scale, relativistic interactions are unable to compete against atomic-scale exchange effects. For example, Heisenberg exchange may exceed 1000 K, whereas typical anisotropies are less than 1 K. However, electrostatic and relativistic contribution become comparable on length scales of order ao/α = 7.25 nm. In addition to the range of interactions, there is the question of interference with structural length scales. For example, Bloch wave functions, which form the basis for the band-structure theory of itinerant magnetism, require infinite crystals with perfect periodicity.

How does nanostructuring interfere with this requirement? Similarly, from a thermodynamic point of view, ferromagnetism is limited to infinite crystals. It fact, the spontaneous magnetization of any finite magnet is zero, because thermal fluctuations cause the magnetization to average. This leads to the next consideration, the dependence of equilibration or averaging times on structural length scales. There is a fundamental distinction between intrinsic and extrinsic properties. Examples of *intrinsic* properties arc the spontaneous magnetization Ms, the Curie temperature Tc, and the anisotropy Kl.

Intrinsic properties describe perfect crystals or surfaces, but their physical origin is atomic and involves quantum phenomena such as exchange, crystal-field interaction, interatomic hopping, and spin-orbit coupling. Intrinsic properties tend to approach their bulk values on fairly small length scales. For example, 'long-range' hermodynamic fluctuations, as involved in the realization of the Curie temperature, and deviations from the Bloch character of metallic wave functions yield only small corrections when the size of the magnetic particle exceeds about 1 nm.

The dynamics is characterized by fast equilibration times which means that intrinsic properties can be treated by equilibrium statistical mechanics. This makes it possible to treat intrinsic properties as local parameters. For example, *M*s(r*)* and *K*1 *(*r*)* reflect the local chemistry, and the unit vector n*(*r*)* of the easy magnetization direction corresponds to the local c-axis orientation of the crystallites. *Extrinsic* or hysteretic magnetic properties, such as the coercivity *H*c and the remanence *M*r, reflect the magnet's real-structure. For example, the coercivity of technical iron doubles by adding 0.01 wt.% nitrogen.

Such small concentrations have little effect on the intrinsic properties but lead to inhomogeneous lattice strains that affect the propagation of magnetic domain walls and explain the observed coercivity increase. The hysteretic character of extrinsic properties means that equilibration times may be very long. At room temperature, the switching of a single atomic moment is a frequent event, but the thermally activated switching of nanoscale cooperative units, such as domain-wall segments, is very rare. This is the thermodynamic origin of hysteresis, enabling us to build permanent magnets and to store information on magnetic disks. This chapter investigates how intrinsic and extrinsic properties are affected by nanostructuring. Emphasis is on model calculations and analytical approximations, as contrasted to Chapter 2 by Kashyap *et al.* which focuses on numerical calculations, where emphasis is on micromagnetic models and simulations.

MESOSCOPIC MAGNETISM

Hysteresis of Magnetic Nanostructures

Magnetic anisotropy yields easy magnetization directions corresponding to local energy minima and energy barriers that separate the easy directions. On an atomic scale, the barriers are easily overcome by thermal fluctuations, but on nanoscale or macroscopic length scales the excitations are usually too weak to overcome the barriers.

This is observed as *magnetic hysteresis*. Zeeman and selfinteraction (demagnetization) magnetic fields, interatomic exchange, and magnetic anisotropy all contribute to the rotation, which occurs on a mesoscopic scale and has been known historically as micromagnetism, although nanomagnetism would be a better name to characterize the involved length scales. Magnetic nanostructures exhibit a particularly rich extrinsic behavior, but even traditional 'microstructured' magnets exploit nanometer-scale features for performance optimization.

For example, the best room-temperature permanent magnets are now made from Nd-Fe-B, but as-cast samples with the correct stoichiometry exhibit a disappointingly low coercivity unless the grain-boundary structure is optmized by a specific heat treatment. This means that magnetic domains are separated by domain walls (dotted lines) whose motion is impeded by real-structure defects or 'pinning centers' in the bulk or at the surface. Aside from a few basic hysteresis mechanisms, such as pinning, coherent rotation, curling, and localized nucleation, there exist many variations and combinations. The reason is the real-structure dependence of magnetic hysteresis, which makes it necessary to consider each material or each class of materials separately.

Micromagnetic Free Energy

A key theoretical problem is to derive magnetization curves by simulating or modeling the magnet's nanostructure. This requires the determination of the local magnetization M(r), from which the hysteresis loop is obtained by averaging. The large strength of the intra-atomic exchange means that typical magnetization changes in magnetic solids are caused by moment rotations rather than by changes in the moments' magnitude. The result is the complicated nonlinear, nonequilibrium, and nonlocal problem of hysteresis. In addition, there is a strong real-structure influence. Properties related to hysteresis (extrinsic properties) are also known as micromagnetic properties, but this term is somewhat unfortunate, because most micromagnetic phenomena are nanoscale.

Hysteresis problems are usually treated on a continuum level. Narrow-wall phenomena, which have been studied for example in rare-earth cobalt permanent magnets and at grain boundaries, involve individual atoms and atomic planes and lead to comparatively small corrections to the extrinsic behavior (§4.2). Furthermore, in contrast to the intrinsic phenomena considered in Section 2, which affect the spontaneous magnetization $M\text{s} = |\mathbf{M}|$, micromagnetic phenomena are realized by local rotations of the magnetization vector. This is because typical micromagnetic energies are much smaller than the quantum-mechanical energy contributions that establish Ms. To explain the hysteresis loop of magnetic materials one needs to trace the local magnetization M(r) = *M*ss(r) as a function of the applied field *H*. This is achieved by considering the free-energy functional

Here *M*s (r) is the spontaneous magnetization, *K*1 (r) is the first uniaxial anisotropy constant, *A*(r) denotes the exchange stiffness, and n(r) is the unit vector of the local anisotropy direction. H is the external magnetic field, and Hd is the magneto static self-interaction field. The latter can be written as The *free*-energy character of *F* reflects the intrinsic or equilibrium temperature dependence of the parameters *A*, *K*1, and *M*s. Furthermore, these parameters are local parameters, because they depend on chemistry, crystal structure, and crystallite orientation.

Depending on the considered system, additional terms must be added to the micromagnetic equation.

Magnetization Dynamics

Magnetization processes are generally time-dependent, even if the external magnetic field is kept constant. For example, freshly magnetized permanent magnets lose a small fraction of their magnetization during the first few hours.

In nanomagnets, the oscillation and relaxation times vary from less than one nanosecond to millions of years.

Atomic processes are very fast, so that intrinsic properties obey equilibrium statistics. An intermediate regime is characterized by typical magnetostatic and anisotropy energies per atom, about 0.1 meV, which correspond to times of order *τ*0 0.1 ns. Examples are ferromagnetic resonance and related precession and damping phenomena. When energy barriers are involved, thermal excitations lead to a relatively slow relaxation governed by the Boltzmann-Arrhenius law.

This relation, known as the Néel-Brown law, goes back to the 1930s. Due to the exponential energy dependence, extrinsic equilibration times vary over many orders of magnitudes, from nanoseconds or milliseconds in superparamagnetic particles to decades in permanent magnets and recording media and millions of years in magnetic rocks.

Fundamental Equations

The time-dependent Schrodinger equation can, in principle, be used to predict the evolution of any physical system, but this method is not feasible in practice. First, the deterministic character of the Schrodinger equation forbids irreversible processes. Second, the many-body character of the Schrodinger equation, and the large number of degrees of freedom, such as lattice vibrations, complicate the description of real magnetic systems. To make meaningful predictions about the relevant magnetic degrees of freedom, such as the position of a domain wall, one must treat the irrelevant atomic degrees of freedom as a heat bath, thereby introducing irreversibility.

CASE STUDIES

Magnetic Localization and Cooperativity

From an atomic point of view, all nanostructures are cooperative, because interatomic exchange ensures well-defined local spin correlations. In other words, nanoscale magnetization processes involve blocks of spins rather than individual spins. The sizes of the correlated regions are nanoscale, in agreement with the analysis in §2.3, but vary from system to system. The coherent-rotation and curling modes are delocalized, that is, they extend throughout the magnet. Delocalized magnetization states are favorable from the point of view of interatomic exchange, because the magnetization gradient is small.

However, exchange is not the only consideration, because local variations of the magnetization cost some exchange energy but may be favorable from the point of view of local anisotropy inhomogenities.

The localization length and the coercivity reduction depend on the ratio of exchange and anisotropy energies. The stronger the anisotropy inhomogenity, the smaller the localization length L, and for a soft inclusion in a very hard matrix, as considered in §2.3, the localization length reduces to the size of the inclusion. For zero disorder, that is, for perfect ellipsoids of revolution, the localization length goes to infinity and the reversal degenerates into coherent rotation.

Cooperative effects are of great importance in advanced technology. For example, in high-density magnetic recording media they lead to the formation of interaction domains, which may improve the thermal stability but reduce the storage density. In permanent magnets the vanishing of the two-phase shoulders in hysteresis loops can be considered as a cooperative effect, as is the above discussed low coercivity of soft-magnetic random-anisotropy magnets. Similar arguments apply to finite-temperature phenomena. On a local scale, the spin alignment remains largely ferromagnetic but thermal excitations and local perturbations break the cooperativity on a nanoscale.

Weakly interacting particles and grains are noncooperative. In this limit, one can use micromagnetic mean-field theories, the Preisach model, and approaches based on Wohlfarth's remanence relation, such as Henkel, delta-M, and delta-H plots. When the interactions exceed a certain threshold, the behavior of the magnet changes from noncooperative to cooperative. In this regime, the above-mentioned approaches are no longer applicable.

For example, the exchange field of strongly coupled small particles exceeds 100 T. In the mean-field approximation, this field adds to the external field and yields an unphysically high coercivity. In reality, the strong exchange field does not translate into a high coercivity, and cooperativity may even reduce the coercivity. This is because two strongly interacting particles behave like one particle, and there is no point in adding any strong internal interaction to the external field.

A rough criterion for the applicability of interaction-field models is obtained from the slope of the hysteresis loop at coercivity, $\chi c = dM(Hc)/dH$. When $Ms/\chi c$ is smaller than the interaction field, then the behavior of the magnet is governed by cooperative effects.

Narrow-Wall and Grain-Boundary Effects

Typical domain walls are smooth and extend over many interatomic distances. However, deviations from this continuum picture occur in very hard materials (narrow walls), at grain boundaries and in the case of geometrical constraints. Narrow-wall phenomena, which have been studied for example in rare-earth cobalt permanent magnets and at grain boundaries, involve individual atoms and atomic planes and lead to comparatively small corrections to the extrinsic behavior.

The nanoscale spin structure at grain boundaries and nanojunctions affects the performance of permanent magnets, magnetic recording media, soft magnets, and structures for spin electronics. In permanent magnets, it is often necessary to maximize or minimize the intergranular exchange, depending on the desired reversal mechanism. Strong intergranular exchange reduces the coercivity of isotropic soft-magnetic nanostructures but is undesired in magnetic recording, where it negatively affects the storage density, and the resistance of spin-electronic structures depends on the local magnetization at interfaces and at junctions. Some aspects of grain-boundary magnetism were anticipated long ago, but polycrystalline granular interfaces and constrained domain walls was first investigated have been investigated only recently.

CONCLUDING REMARKS

From a scientific point of view, nanomagnetic effects are intermediate between atomic-scale magnetism and macroscopic phenomena but cannot be reduced to a mixture of the two limits. The main reason is the competing involvement of relativistic interactions, which give rise to an additional length scale of $ao/\alpha = 7.25$ nm. Intrinsic roperties, such as spontaneous magnetization, Curie temperature, and magnetocrystalline anisotropy, reflect comparatively strong quantum-mechanical and spin-orbit interactions. They are realized on atomic length scales and therefore well-defined for nanostructures. However, surface and interface contributions such as magnetic interface an-isotropy may be strong and outweigh bulk contributions on a nanoscale. Extrinsic properties, such as coercivity, are realized on larger length scales and exhibit a pronounced real-structure dependence.

A particularly subtle property is the Curie temperature, which describes the onset of ferromagnetism and involves long-range thermodynamic fluctuations. In a strict sense, the Curie temperature of finite bodies is zero, but the nonequilibrium character of practically encountered magnetic phenomena and the low energy differences associated with long-range thermodynamic fluctuations mean that particles or grains larger than about 1 nm., are virtually indistinguishable from true ferromagnets. For this reason, nanostructuring cannot be used to improve the Curie temperature.

It is important to distinguish between equilibrium and nonequilibrium extrinsic properties. The equilibrium behavior of magnetic nanostructures, as epitomized by the critical single-domain radius, is largely irrelevant to hysteresis. For example, in highly anisotropic rare-earth transition-metal permanent magnets, the critical single-domain size is of the order of 1 ìm, but this does not mean that the magnetization reversal in such a particle is Stoner Wohlfarth like. In fact, the reversal is localized, and typical coercivities are realized on a length scales of a few nanometers.

From a structural point of view, surfaces, interfaces, and bottlenecks (junctions) have a strong impact on nanomagnetism. One issue is that exchange at grain boundaries affects the coupling between nanograins and, indirectly, the extrinsic properties of the structures. Reduced grain-boundary exchange leads to a quasi-discontinuity of the magnetization. By contrast, anisotropy changes in the grain-boundary region have no major effect on the spin structure.

Intrinsic properties correspond to a very fast equilibrium, whereas the equilibration times of extrinsic properties cover a broad range, from a about one nanosecond in soft-magnetic resonance experiments and to millions of years in magnetic rocks. The slowest mechanism is thermally activated jumps over nanomagnetic energy barriers.

This results in small low-temperature magnetic-viscosity corrections to the leading intrinsic contribution. Since the phase-space trajectories responsible for thermally A activated magnetization reversal are very close to static trajectories, 'giant' thermodynamic fluctuations involving arbitrary modes can safely be ignored. In conclusion, magnetic nanostructures exhibit various scientifically interesting and technologically important deviations from bulk and thin-film magnets.

For example, there is a resonant quantum entanglement of the low- lying states of coupled hard-magnetic dots, which may be exploited in future quantum-information processing above 4.2 K. Some others examples, mentioned or elaborated here and throughout this book, are the energy-product enhancement in nanostructured two-phase magnets whose hard-magnetic performance is improved by adding a soft-magnetic phase, multilayered and granular spin-valve structures, nanostructured soft magnets, and magnetic recording media. The principles and mechanisms outlined in this chapter make it possible to realize properties not achievable in single-phase bulk and thin-film materials, and form a basis for future experimental, technological and theoretical developments in the fascinating field of nanomagnetism.

5

Advanced Magnetic Circuit

An electromagnetic pickup for a stringed musical instrument incorporating a magnetic structure placed between two elongated collinear coils of opposing polarity. The magnetic structure includes a number of ferromagnetic polepieces that extend through the collinear coils and two ferromagnetic moderator bars adjacent to and on opposite sides of the polepieces (within the area between the two coils) that are charged with one polarity by high coercive rare earth high coercive permanent magnets attached to the outer surface of the moderator bars.

Claims:

1. A magnetic pickup for stringed musical instruments having ferromagnetic strings, the pickup comprising: a pair of axially spaced coils to have their axis extended perpendicular to the length of the strings: a plurality of ferromagnetic poles extending through the first and second coils; first and second coils are separated by first and second ferromagnetic moderator bars adjacent to each side of the ferromagnetic poles; and high coercive permanent magnets are affixed to the sides of the first and second ferromagnetic bars opposite the plurality of ferromagnetic poles.

2. The pickup according to claim 1 in which two coils are connected, either in series or parallel mode, in an out of phase relationship to each other.
3. The pickup assembly according to claim 1 in which the coils are respectively disposed on a pair of bobbins, the bobbins and ferromagnetic bars being held together by the plurality of ferromagnetic poles that are press fit into the bobbins.
4. The pickup according to claim 1 in which the coils, and the poles are elongated in a direction transverse to the string length as to be responsive to a plurality of parallel strings.
5. The pickup according to claim 1 wherein the separation between the coils is of a greater dimension than the height of each of the coils and the total height of the pickup does not exceed 0.900 inches.
6. The magnetic pickup according to claim 1 wherein the high coercive permanent magnets have approximately half the mass of the ferromagnetic bars.
7. The magnetic pickup according to claim 1 wherein the high coercive permanent magnets are magnetically oriented parallel to the direction of the strings disposed with one magnetic polarity to the ferromagnetic bars which induce the same magnetic polarity into the plurality of ferromagnetic poles and the other magnetic polarity is directed parallel to the strings away from the pickup.
8. The magnetic pickup according to claim 1 in which two the coils are not directly influenced by the dipole function of the high coercive permanent magnets.

Description:

Field of the Invention

This invention relates to an electromagnetic pickup for converting vibrations of a musical instrument string into corresponding electrical signals, for example, a guitar pickup.

Background of the Invention

The traditional single coil magnetic pickup for stringed musical instruments as originally designed and offered by Fender Musical Instruments Corp. consists of a coil form with alnico magnets as its core that is wound with numerous turns of copper wire. This design set the standard for reproducing what is commonly known and recognized today as the Fender sound. Inherent to this original work are certain design choices that have been well documented and in some cases addressed to various degrees by other pickup designs. To some degree, many early pickup designs have been susceptibility to external electromagnetic radiation and the magnetic interference with the natural vibrations of the strings.

To address the electromagnetic interference, hum canceling coils, also known as humbucking coils, have been used in pickup designs. Hum canceling coils have been known since 1825 when Leopoldo Nobili invented the astatic galvanometer; where it is stated: "Two identical coils of N turns are connected in series, in such a way, that the current passes through them in opposite senses to neutralize external magnetic fields". Collinear coil assemblies for electrical musical instruments have been known since the mid 1930s, with U.S. Pat. No. 2,119,584 most likely representing the first such patent. Many early pickup designs, due to the extremely small market demand at the time, were never patented.

Magnetic pickups of various designs have been used in the sound reproduction of stringed musical instruments since 1930. They are generally placed directly under the strings between the bridge and the end of the fingerboard of the instrument. In simplest terms,. the electromagnetic pickup in combination with the vibrating string represents a multi-frequency voltage generator. Magnetic pickups are an important component in what makes up the sound characteristics of a given amplified electric stringed musical instrument. Certain companies that manufacture these instruments have developed magnetic pickups that have contributed to what has become known over time as their signature sound. The acknowledgement of this sound signature has to be considered in the design of any new instrument and its component parts including the design of the pickup.

Based on changes that had taken place in musical instrument amplification (for example, digital sound recording equipment; computer interface for stringed musical instruments; changes in playing techniques and styles of music), Fender requested, in 1996, the development of a high performance electromagnetic pickup.

The pickup had to meet the following demands: (1) to not exceed the dimensions of Fender's traditional single coil pickups; (2) to generate a higher output voltage than their traditional single coil pickups; (3) to reproduce the legendary signature sound of their traditional single coil pickups; (4) to increase sensitivity to minute amplification adjustments allowing the player to select from a wider tonal variety; (5) to reduce the magnetic force directed towards the strings so as to minimize magnetic interference with the vibrating strings; (6) to neutralize the interference of external electromagnetic radiation; and (7) to be manufactured cost effectively.

These demands presented several major problems. To increase the output requires either stronger permanent magnets or more turns of wire. Stronger permanent magnets, due to their high coercive force, have negative solenoidal qualities. When placed close to the core of a solenoid they will cause magnetic saturation of the core resulting in a dramatic loss of their relative permeability and in addition will cause a strong magnetic interference with the vibrating string. More turns of wire would not only increase the impedance, resulting in a shift to undesired frequencies, but would also require a larger coil dimension.

It was also important to consider that a vibrating string does not produce a pure tone that can be explained with the single curve of a sine wave. The tone of a vibrating string consists of several different sine waves, resulting in a complex waveform that can be found by adding the ordinates of all its component sine waves. This complex waveform represents the Fourier spectrum of a tone, which musicians simply call the "signature sound". The Fourier spectrum depends partially on the position of a pickup in relation to the bridge of the instrument and the position, angle, and force of the attack. However, the most important factor is that the signal generated above the coil is in-phase with the signal generated at the sides of the coil.

A need exists to eliminate the negative functions of magnetic structures used in collinear coil assemblies.

Summary of the Invention

In this invention, the electromagnetic pickup comprises a magnetic structure which minimizes the magnetic effect on the natural vibration of the strings, and primarily provides an equal charge distribution of moderate magnetic intensity focused towards the strings and a secondary charge of much greater magnitude below the active coil parallel to the strings exceeding the width of the pickup. The magnetic structure includes a number of polepieces and two moderator bars. The materials of both bars and the polepieces are of high magnetic susceptibility to increase the self inductance in each of the coils, while reducing the negative mutual inductance inherent in collinear coils of reversed polarity. The moderator bars are charged with high coercive permanent magnets having about half the mass of the moderator bars, which are placed on the outer surface of the bars, i.e., the outer perimeter of the pickup. The combination of mass and placement of the high coercive permanent magnets eliminates any negative interferences. The magnetic structure is designed to enhance the performance of a pickup with two elongated collinear coils of opposing polarity to neutralize external electromagnetic radiation.

An additional objective in the design of this invention is that it could be manufactured cost effectively.

Other advantages, features and additional objects of the present invention will become manifest to those versed in the art upon making reference to the detailed description and the accompanying drawings in which a preferred structural embodiment incorporating the principles of the present invention is shown by way of illustrative example.

The present invention is described in one or more embodiments in the following description with reference to the Figures, in which like numerals represent the same or similar elements. While the invention is described in terms of the best mode for achieving the invention's objectives, it will be appreciated by those skilled in the art that it is intended to cover alternatives, modifications, and equivalents as may be included within the spirit and scope of the invention as defined by the appended claims and their equivalents as supported by the following disclosure and drawings.

The present invention is incorporated in an electric guitar, typically but not limited to a solid-body electric guitar. The word "guitar", as employed in the present specification and claims, denotes any electric guitar, electric bass guitar, or any electric string musical instrument incorporating electromagnetic pickups.

The pickup shown in the drawings is for six-string guitars. However, the number of strings and the number of polepieces per pickup may vary with the design of the guitar.

The bottom flange plate of the lower bobbin is extended on each end outside the perimeter of the core assembly with a hole in the extended area on each side used when affixing the pickup to the guitar. In addition, there is a printed circuit board or array on the bottom surface of the flange plate that is used to connect the pickup to the other electronic components of the guitar.

Each flange plate has a plurality of circular holes that correspond to the number of polepieces used for that particular pickup. These holes are to a dimension corresponding to the outside diameter of the polepieces so as to allow the assembly to be pressed together. An additional smaller diameter hole located in the center of each flange plate serves a dual purpose. It is used in mounting the core assembly to the coil winding machine and it allows for the release of air during the waxing process, allowing the wax to fully permeate the coil so as to avoid microphonic squealing.

The illustrated ferromagnetic pole pieces are of sufficient length to extend fully through the upper and lower bobbin sections.

To further describe the pickup shown in drawing FIGS. 1-5, it comprises an upper bobbin 18, a lower bobbin 19, six ferromagnetic pole pieces 6a, 6b, 6c, 6d, 6e, and 6f, two ferromagnetic moderator bars 7a and 7b, four high coercive permanent magnets 8a, 8b, 8c, and 8d, upper and lower coils or windings of wire 5a and 5b, and two terminal plates 9a and 9b.

Flange plates 1, 2, 3 and 4 are made from an electrically insulating material, preferably fiberglass. Upper bobbin 18 has top flange plate 1 and bottom flange plate 2 mounted parallel to each other, spaced to a predetermined width by two tubular spacers 10a and 10b, to assemble a bobbin onto which coil 5a is wound. Lower bobbin 19 has top flange plate 3 and bottom flange plate 4 mounted parallel to each other, spaced to a predetermined width by two tubular spacers 10c and 10d, to assemble a bobbin onto which coil 5b is wound, in reverse polarity of upper coil 5a. The coils 5a and 5b are parallel to each other. Flange plates 2 and 3 have slots to protect wire that is extended down to flange plate 4. Flange plate 4 is extended on each end outside the perimeter of the core assembly with a hole in the extended area on each side for mounting purposes. In addition, there is a printed circuit array on the bottom surface of flange plate 4 that is used in connecting to the other electronic components of the guitar.

The six ferromagnetic pole pieces 6a, 6b, 6c, 6d, 6e, and 6f are mounted parallel to each other through the holes in the flange plates 1 and 2 of upper bobbin 18, and flange plates 3 and 4 of lower bobbin 19, as shown, using friction to hold the bobbin assembly together. In addition, ferromagnetic pole piece 6apasses through spacers lob and 10c, while ferromagnetic pole piece 6f passes through spacers 10a and 10d.

The ferromagnetic moderator bars 7a and 7b are mounted between and perpendicular to flange plates 3 and 4; adjacent to the six ferromagnetic polepieces 6a, 6b, 6c, 6d, 6e, and 6f.

The four high coercive permanent magnets 8a, 8b, 8cand 8d are mounted between and perpendicular to the flange plates 3 and 4; affixed directly to the outer surface of the ferromagnetic moderator bars 7a and 7b. High coercive permanent magnets 8a and 8b are affixed to the ferromagnetic moderator bar 7a, and high coercive permanent magnets 8c and 8d are affixed to the ferromagnetic moderator bar 7b.

The magnetic north pole of the four high coercive permanent magnets 8a, 8b, 8c, and 8d is on the side facing the outer surface of moderator bars 7a and 7b.

Electrical connection points 12 and 16 on the bottom surface of flange plate 4, and electrical connection points on the two terminal plates 9a and 9b, allow that the coils can be connected in either series or parallel modes in opposite polarities. In both the series and parallel configurations, connection points 13 and 15 represent the resulting connections of both coils to allow reversed polarity wirings in multi-pickup combinations. Connection point 14 represents the common ground for the magnetic poles via the PC board (flange plate 4).

High coercive permanent magnets have negative solenoidal qualities. When placed close to the core of a solenoid they cause a high saturation of the core resulting in a dramatic loss of the cores relative permeability and causing an increase of the effective resistance of the coil. The other factor is that the increase of magnetic force of the polepieces towards the strings will hinder the strings from vibrating evenly, resulting in false overtones and loss of sustain.

In this invention the ferromagnetic moderator bars serve two different functions: (A) magnetic and (B) solenoidal. In function A, the moderator bars equally distribute a reduced magnetic charge, induced by the high coercive source magnets, to the ferromagnetic pole pieces, thus minimizing the magnetic effect on the natural vibration of the strings, and since the ferromagnetic bars have a greater mass and greater dimension than the high coercive permanent source magnets, to avoid magnetic fringing. In function B, the moderator bars increase the self inductance of the individual coils, and reduce the leakage inductance between the coils, to predetermined magnitudes to cancel eddy currents.

In this invention the high coercive permanent magnets 8a, 8b, 8c and 8d have about half the mass of the ferromagnetic moderator bars 7a and 7b, and are placed adjacent to the ferromagnetic moderator bars 7a and 7b close to the perimeter of the pickup, without making direct contact with ferromagnetic polepieces 6a, 6b, 6c, 6d, 6eand 6f, and are not under, next to or inside the core assembly thus meeting the criteria as described above.

While placing the high coercive permanent magnets 8a, 8b, 8c, and 8d on the outside surface of the ferromagnetic moderator bars 7a and 7b reduces the magnetic force transmitted to the polepieces 6a, 6b, 6c, 6d, 6e, and 6f to a determined quantity, the ferromagnetic moderator bars 7a and 7b function as magnetic keepers increasing the magnetic force on the outside surface of the high coercive permanent magnets 8a, 8b, 8c, and 8d creating a strong magnet field that is parallel to the string 15. The parallel field, while it does not interrupt the natural vibration of the string 15, does extend the width of the magnetic field beyond the physical dimension of the pickup itself.

While one or more embodiments of the present invention have been illustrated in detail, the skilled artisan will appreciate that modifications and adaptations to those embodiments may be made without departing from the scope of the present invention as set forth in the following claims.

RESEARCH MAY ADVANCE TECHNOLOGIES FOR ADVANCED MAGNETIC-RECORDING DEVICES

A new method for exploring the secrets of Mother Nature's frustrations has been developed by a team of physicists lead by Penn State University professors Peter Schiffer, Vincent Crespi, and Nitin Samarth. The research, which will be published this week in the journal Nature, is is an important contribution to the study of complex interacting systems, and it also could contribute to technologies for advanced magnetic-recording devices.

"We all would prefer to have less personal experience with frustration, but the state of frustration also is an important factor in the way many systems in nature work," explains Schiffer, who is a professor of physics at Penn State. "Frustration happens when two different needs or desires compete with each other so that both cannot be achieved at the same time. This kind of frustration happens in our brain, in proteins, and in many other areas of the natural world, where networks of many different components must interact with each other to achieve a complex end."

Schiffer explains, for example, that neural networks, which allow the brain to function, and protein molecules, which allow living matter to function, consist of thousands to millions of interacting components, and that a crucial element of these interactions is that they often are "frustrated." "When two different and competing signals are sent in the brain, the brain needs to choose which signal will dominate in order to take a particular action," Schiffer says. "Frustration happens even in a simple substance such as ice, which consists of only hydrogen and oxygen atoms, because there are competing forces on the hydrogen atoms pushing them between different positions relative to their neighboring oxygen atoms," he explains.

Understanding the consequences of frustration in an extremely complex system like the brain is very difficult, so there is a great deal of research interest in studying simpler frustrated systems, like ice, in order to obtain a basic understanding of the nature of frustration. One group of such systems are materials in which some of the individual atoms have "magnetic moments," meaning that each atoms is like a tiny bar magnet or compass needle. If a material has these atoms arranged in certain ways, the interactions among groups of magnetic atoms compete with each other, which leads to a state of frustration. These "frustrated magnetic materials" are perhaps the cleanest systems in which frustration can be studied and have been the subject of intense research.

"The direction along which the magnetic moment of these magnetic atoms is pointed is determined by interactions with the other magnetic atoms in the material," Schiffer explains; "however, it has been almost impossible to look at the magnetic states of individual atoms, and existing chemical-synthesis techniques do not permit the strength of the interactions within these materials to be easily tuned." As a result, physicists have been able to study only the collective behavior of a group of frustrated magnetic atoms—not the specific behavior of the individual atoms.

Now, Schiffer and his colleagues have developed a new method to study the subtleties of frustration, which involves using "electron beam lithography" to build a magnetically frustrated material by sculpting arrays of hundreds of thousands of microscopic bar magnets, each only a few millionths of an inch in size. The samples were fabricated at the Penn State Nanofabrication Facility by Penn State graduate student Ruifang Wang. "Rather than relying on chemically synthesized materials whose magnetic atoms are pre-arranged, we decided to make our own frustrated system," Samarth says.

"We tailored our system to mimic the magnetic structure of a specific set of materials that we wished to study, in which the frustrated magnetic atoms are arranged so that they behave in a manner exactly analogous to the hydrogen atoms in ice. Ruifang put in a truly heroic effort in getting the experiment to work." This set of materials is called "spin ice" since "spin" is another way of denoting the magnetism of an individual atom.

Because the systems made by Schiffer and his colleagues were created intentionally with a particular size and arrangement of individual magnets, the researchers were able to arrange the system so that the interactions between the magnets were frustrated in the same manner as in spin-ice materials. "Using a special microscope, we could see the direction in which each magnetic moment was pointing in this artificial spin ice, something which is impossible with either chemically synthesized frustrated magnetic materials or with ordinary ice," Schiffer says. "We also could change the spacing of the arrays, which allows us to tune the strength of the frustrated interactions."

The researchers say this work is an important step forward in the study of the nature of frustration in large networks. "Using such fabricated arrays, it now is possible to engineer frustrated systems to alter the strength of interactions, the geometry of the lattice, the type and number of defects, and other properties that impact the nature of frustration," Schiffer explains. These systems also allow scientists to probe the state of individual elements within a frustrated system. The interacting magnets also are relevant to modern magnetic-recording technology, which relies on increasingly tiny magnetic structures in order to fit more and more information onto smaller and smaller hard drives.

Modelling has played a very important role in the development and improvement of new materials for applications. Our knowledge of nanomaterials such as quantum dots, nanostructures, molecular assembly, proteins or DNA molecules, is still at a very begining stage. Modelling and simulation techniques affect all stages in the development and improvement of new materials, from the initial formation of concepts to synthesis, characterization and properties. We think new modeling strategies for complex material systems are needed. At present time it still seems unlikely that we will be able to treat all of the interesting problems by *ab initio* methods.

6

A New Startegy

Even if this were to become possible, the analysis of the enormous amount of data generated from computer simulations would itself require massive computing power and a comprehensive modelling strategy.

Such modelling strategies should focus on the multi-scale and multiphysics nature of emergent problem in area. We gently introduce here the Artificial Neural Network (ANN). In our first model we intend to use for the input data e.g. the chemical composition, thermal treatments and some structure data. As output data: AHE, and magneto-resistance effect of some of our new magnetic Co-based alloys.

INTRODUCTION

Beside the exciting new capabilities and opportunities, nanostructured materials cause some challenges that were not present in the field of *traditional* materials. While trial-and-error engineering was a successful approach for most of the traditional materials, one can show why this is difficult in the nanoscale materials domain, where fundamental understanding on the atomic level is required for the successful materials design. In this area, computer modeling poses new chances, for targeted design - when developing new magnetic nanomaterials.

Characterization

For traditional materials, the microstructure and the composition - weight fractions of the different elements in the mixture, determine their properties. For nanostructured materials, the exact composition (what atomic species are present), and the configuration (where the atoms and sometimes even electrons are exactly located on the atomic scale), determine the materials properties. Conventional tools such as light microscopes, X-ray diffraction, mass spectrometers etc., cannot provide such information. Combinations of atomic-level modeling and atomic-resolution characterization e.g. analytical electronmicroscopy gave us hopes in this area so far.

Fabrication - process Modeling

For traditional materials average properties such as concentrations in continuum or phase-field modeling and mostly equilibrium thermodynamics - phase diagrams, diffusivities are sufficient to model the evolution of a material during processing. For nanostructured materials, single atoms need to be traced during the processing in order to predict the final structure and its properties. However, traditional atomic-level techniques such as molecular dynamics are only capable of following a few atoms reliably for few nanoseconds, not long enough for most processing steps. The recent development of new techniques such as transition-state-theory based methods, accelerated dynamics and physics-based Monte Carlo methods get around this problem and enable nanoscale process modeling.

Structure-property Relationship

For traditional materials, the properties are determined by microstructure and composition. The structure-property relationships are in many cases still poorly understood from a fundamental point of view, but vast experience and theory exists which tells us how to change the structure in order to improve the properties. For nanostructured materials, the structure-property relationships can only be found on the atomic level and experience is missing. As examples there are the scaling relationships between dimensions and electron transport in nanoelectronics devices, which strongly differ from those of the conventional devices. The theory that is currently being developed is less heuristic and more fundamental than for traditional materials, which opens the chance of designing and understanding materials more reliably by using computer simulations.

Nanomagnetic Materials Design

For traditional materials, revolutionary new materials are in most cases discovered by accident. Most of the *design* time is spent on incremental improvement of a known type of materials, e.g. alloying it with small amounts of impurities. For nanostructured materials, computational design of hypothetical new materials is possible, including the study of stability and the prediction of properties. In combination with a mature materials synthesis facility completely new materials can be virtually designed and made-up. Nanomagnetism research involves investigating the basic magnetic, magneto-optical, galvanomagnetic, magnetotransport phenomena associated with reduced dimensionality.

It has long been known that confinement and finite-size effects alter magnetic properties. Recently, it has been shown that these finite-size effects can be connected for new magnetic applications. Thus, it is vital to understand the magnetic behavior of individual building blocks, which eventually will be combined into more complex structures for new functionalities. In addition to technological relevance, other exciting basic science questions remain to be answered, which arise from the interplay of finite size and shape induced modifications of magnetic behavior combined with interactions across interfaces between magnetically different materials. Nanopowders, nanowires, thin films, (Fe, Co, NiCo, CoFeSiB) and magnetic nanomaterials recently prepared in our laboratories, now need our theoretical consideration.

Our research is focused on a large extent on developing a robust IT method to solve the *prediction problem* for a *nice* relationship structure-property concerning our magnetic nanostructures. We try to obtain at the beginning, a new model for the characterization of some materials in the class: FeCrSiBMo and CoFeSiB alloys. They were studied with other methods for some interesting properties like: Hall Effect, magnetoresistance effect on the influence of the composition and of the thermal treatments for the magnetic properties of this alloys. Now we don't have analytical solution for them.

Once we have a crisp definition, the next step is to select the input variables and the desired responses. So we introduce the new strategy, we bring in the *Artificial Neural Network* (ANN). The IT method used is based on an adapted Neural Nwtwork with the *Neuro Dimension* platform. We notice some major applications of ANN's as follow: Pattern classifiers: The necessity of a data set in classes is a very common problem in information processing. We find it in quality control, financial forecasting, laboratory research, targeted marketing, bankruptcy prediction, optical character recognition, etc.

ANNs of the feedforward type, normally called multilayer perceptrons (MLPs) have been applied in these areas because they are excellent functional mappers - these problems can be formulated as finding a good input-output map. Associative memories: Human memory principles seem to be of this type. In an associative memory, inputs are grouped by common characteristics, or facts are related. Networks implementing associative memory belong generally to the recurrent topology type, such as the Hopfield network or the bi-directional associative memory. However, there are simpler associative memories such as the linear or nonlinear feedforward associative memories.

Feature Extractors

This is also an important building block for intelligent systems. An important aspect of information processing is simply to use relevant information, and discard the rest. ANNs can be used as principal component analyzers, vector quantizers, or clustering networks. They are based on the idea of competition, and normally have very simple one-layer topologies. Dynamic networks: A number of important engineering applications require the processing of time-varying information, such as speech recognition, adaptive control, time series prediction, financial forecasting, radar/sonar signature recognition and nonlinear dynamic modeling. The field of neural networks has been studied for almost five decades but has found solid application only in the past fifteen years, and this field is still developing quickly. Today neural networks can be trained to solve problems that are difficult for conventional computers or human beings.

THEORY ON NANOMAGNETIC MATERIALS AND EXPERIMENTAL ASPECTS

One know that in a very simplified atomic orbital model, the moment is created by interaction between electrons on the same atom and the coupling between electrons on diferent atoms. Spins are aligned antiparallel at the smallest distances or volumes, since Pauli' s exclusion principle prevents equal spins from occupying the same status. This reduces the moment of an individual atom and produces antiferromagnetic coupling between atoms. The interaction changes its sign as soon as the wavefunctions cease to overlap strongly. The moment increases and the coupling becomes ferromagnetic. At larger distances the magnetic coupling becomes indirect, using conduction electrons as mediators. It oscillates with a period determined by Fermi wave-vectors, thereby bringing orientation-dependent Fermi surface parameters into the model. More sophisticated treatments, such as local density theory, take the band-like nature of the 3d wavefunctions in metallic ferromagnets into account, which distort the simple atomic picture.

Magnetism reflects the essential band structure. The magnetic moment is given by the difference between the filling of majority and minority spin bands, which is directly related to the ferromagnetic exchange splitting between the bands. Total energy minimization over all occupied band states yields crystal structure and magnetic ordering. Magnetic coupling and magnetic moment are sensitive functions of the interatomic spacing and the atomic volume respectively. Both are dominated by the Coulomb and, or exchange interactions at small distances. Magnetic domains provide a large-scale texture that adds enormous complexity. As a consequence, there are many more magnetic properties than band-structure parameters, and a large number of techniques to measure them.

The variability of an H(M) hysteresis loop reflects a complex domain structure. The situation becomes simpler in nanostructures, which are smaller than the typical extension of a domain wall (0.1-1μm). They often exhibit a square hysteresis loop. Magnetic anisotropy and magnetostriction in thin films demonstrate that magnetic ordering is sensitive to small energy changes that are induced by the presence of an interface and by strain. Magnetization and magnetic coupling are calculable from first principles by band theory.

Actually, the most common approach uses the local density formalism, where the spin-dependent part of the total energy is approximated by an expression derived from an electron gas. Different magnetic structures, such as paramagnetic, ferromagnetic and various antiferromagnetic configurations, exhibit different total energies, and the lowest-energy configuration obtained by local density theory represents the observed magnetic state in most cases. Ferromagnetism occurs in rather few elements, Fe, Co, Ni, Gd and a few rare earths with low Curie temperatures.

The stability of ferromagnetism in these elements can be explained by the Stoner criterion, which takes the density of states at the Fermi level and an atomic exchange integral as input. Stoner criterion explains why Fe, Co and Ni are single for ferromagnetism, several other elements, close to acomplish the criterion, for example Pd. In some thin-film structures these elements are transformed into ferromagnets, i.e. V, Cr, Mn, Mo, Ru, Rh, Pd, Pt. Magnetism can either be induced by exchange coupling to a ferromagnetic substrate or occur spontaneously owing to a higher density of states in a monolayer.

Magnetic anisotropy is a magnetic variable that becomes rather changeable as the bulk symmetry is reduced in small nanostructures. The magnetic energy density E varies quadratically with the angle è of the magnetization with respect to a symmetry axis in a crystal or thin film: E = K cos2, the anisotropy constant, K can be negative or positive. The source of crystalline bulk anisotropy is the spin-orbit interaction. In nanostructures the symmetry of the system is lowered by the existence of a surface or interface, where orbitals perpendicular and parallel to the interface become inequivalent. This symmetry breaking, drive the anisotropy from a fourth-order effect to a second-order effect i.e. in Fe and Ni.

Micromagnetic modelling typically includes a minimization over several energy terms, such as the exchange energy, crystalline anisotropy, the magnetostatic energy of the magnetic moment in the demagnetizing field, and the Zeeman energy of the magnetic moment in an external field. To see what might happen to particle-particle interactions at sublithographic dimensions one can look at the magnetic behaviour of random arrays of magnetic islands. They exhibit distinct magnetic phases, such as perpendicularly magnetized random antiferromagnet (island diameters, less than 3 nm), in-plane superparamagnet (diameters, 3-6 nm) and in-plane ferromagnet (diameters, greater than 6 nm).

Many properties of thin films can be described by three dimensional bands, or by magnetic energies that are proportional to the volume of the film, they are dramatically modified by the influence of the substrate. As the film thickness is reduced, one can ask the question when, and how the transition to two-dimensional behaviour takes place. As the thickness of a thin film decreases, one expects to make a transition from three to two-dimensional behaviour. To quantify this transition, for example to see whether it is abrupt or continuous, we have to find characteristics of three-and two-dimensional magnetism.

Near a magnetic phase transition there exist characteristic power laws for magnetic quantities, such as the magnetization. The Curie temperature is thickness dependent because the reduced number of neighbours in a thin film reduces the overall magnetic coupling. Another signature of two-dimensional behaviour is the transition from *parallel to perpendicular anisotropy* in a thin film. Most of the concepts developed for surface states in planar structures can be extended to one-dimensional wire or stripe structures. Magnetic domain-wall motion in soft magnetic thin films was observed in response to a varying in-plane magnetic field. Domain walls became trapped at nanoscale defects and were then studied using standard TEM imaging.

Several different types of films were studied, including Ni-Fe alloys, Ni-Fe-Co-Si-B amorphous films, and Co-Fe-B amorphous films. Concentrated efforts are now focused in our labs finding innovative and modern solutions to obtain soft and hard nanostructured magnetic materials - magnetic nanocomposites, and also to understand the intrinsic mechanisms governing their specific magnetic behavior. Fe, Co, NiCo and CoFeSiB magnetic nanopowders of 20-200 nm have been prepared by chemical reduction. Due to their specific magnetic behavior, very different compared with their bulk counterparts, the magnetic nanopowders might be successfully used as soft or permanent magnets.

Ni, Co, NiCo, and amorphous Ni 100-x Fe x, Co 100 x Fe x nanowire arrays of 20 to 200 nm in diameter and 1 to 10 μm in length were prepared by electrodeposition. Their magnetic characteristics: coercive field, magnetic hysteresis loops, magnetic domains structure, ga anomagnetic properties, magnetic interactions between nanowires was comparatively studied.

Magnetic nanostructures have been also obtained. Moreover, the nanostructures have been obtained directly during the preparation process, in Fe-(Au,Cu)-B and (Fe,Co,Ni)- Nb-B systems, without any intermediate treatments, in the *as-cast state* in a few families of permanent magnets. Nanocrystalline regions of 10 to 20 nm randomly distributed in the amorphous matrix have been revealed in Fe-(Au,Cu)-B melt-spun ribbons and (Fe,Co,Ni)-(Zr,Nb,Ti,Ta,Mo)-B bulk shaped samples (torroids, bars) prepared by conventional mould-casting techniques.

The IT based on ANN Method

We introduce here the use of Artificial Neural Network (ANN) method. We always use common sense to select the variables that are relevant for the problem. One should seek variables and conditions that appear relevant to the problem being analyzed. One should also seek data that cover a wide spectrum of cases. If the ANN does not see an equilibrated set of cases, its output will be 'biased'. The designing and performing of the new nanocrystalline and amorphous alloys with targeted properties is an actual problem for the ferromagnetic materials. Our model is intented to enhance some new properties of these materials. Because this new method requires a large amount of data – some of them measurable, some other not, colected from experiences made using many different substances and materials with magnetic properties – we will use both data from our laboratory and data already published in reviews.

Unlike more analytically based information processing methods, neural computation effectively explores the information contained within input data, without further assumptions. The methods are based on assumptions about input data ensembles. Artificial intelligence encodes a priori human knowledge with simple if-then rules, performing inference on these rules to reach a conclusion.

Neural networks, *discover* relationships in the input data sets through the iterative presentation of the data and the intrinsic mapping characteristics of neural topologies - *learning*. There are two basic phases in neural network operation.

The *training* or learning phase where data is repeatedly presented to the network, while its weights are updated to obtain a desired response; and the recall or *retrieval phase*, where the trained network with frozen weights is applied to data which were never seen. The learning phase is time consuming due to the iterative nature of searching for the best performance. Once the network was trained, the retrieval phase can be very fast, because processing can be distributed. We notice that a lot of real world problems fall in this category, ranging from classification of irregular patterns, forecasting, noise reduction and control applications. Humans solve problems in a similar way.

They observe events to extract patterns and then make generalizations based on their observations. Neural networks were used for both regression and classification. In regression the outputs represent some desired, continuously valued transformation of the input patterns. In classification, the objective is to assign the input patterns to one of several categories or classes, usually represented by outputs restricted to lie in the range from 0 to 1, so that they represent the probability of class membership. For regression, it can be shown that a single hidden layer Multilayer Perceptron (MLP) can learn any desired continuous input-output mapping if there are sufficient numbers of axons in the hidden layer(s). For classification, Multilayer Perceptrons can learn the Bayesian posterior probability of correct classification. This means that the neural network takes into account the relative frequency of occurrence of the classes, giving more weight to frequently occurring classes. Neural networks have been trained to perform this complex functions for identification and classification.

RESULTS AND DISCUSSION

We advance a new method to characterize the magnetic properties of nanomagnetic materials. Our work purposes: 1) To develop the underlaying theory and a general simulation approach to handle nanomagnetic materials; 2) To develop the large-scale computing approach necessary to simulate nano scale magnetic materials and the distributed infrastructure to make easy the rapid development of new simulation programs and their use directly by experimentalists; 3) To work closely with experimentalists in order to verify models and to explore new nanoscale magnetic materials; 4) High-performance computing aspects of physics-based modeling of nanomagnetic materials.

The simulation of a neural network requires the orchestration of many pieces. When we run the simulation we start by checking if the data was correctly fed into the network by placing probes on the input sources. Another important aspect is to check if the learning rates are sufficiently low to avoid divergence. Divergence will usually occur during the beginning of the training. We place a matrix viewer on the first synapse to see if the weights are changing. Observing a steady decrease of the cost is the best overall indicator that everything is progressing well. Most of the time, the first check is to see if the cost is within what we think is appropriate for the application. All the information gathered by the network from the input data is contained in the weights. So, we should save the weights, along with the topology.

The research is intended to set the foundation for more long-term work on on-line learning and construction techniques for modelling. Training is central to the use of neural networks for modelling since this is still a computationally intensive and time-consuming task, particularly for large-scale problems. Very often the success or failure of a neural network based control strategy depends on whether or not an acceptable model can be found in a reasonable amount of time. It was therefore vital for us to chose a fast, efficient and reliable training method applying this technology in our practical purpose.

Data Structures

The format of input data structures affects the simulation of networks. For concurrent vectors, the order is not important, and if we have a number of networks running in parallel, we could present one input vector to each of the networks. For sequential vectors, the order in which the vectors appear is important. In our model we use for the input data the following: the chemical composition, the thermal treatments and some structure data. As output data Hall Effect (AHE), or magnetoresistance effect of one of our nanomagnetic alloys.

Training Styles. In incremental training the weights and biases of the network are updated each time an input was presented to the network. Learning Rules. We define a learning rule as a procedure for modifying the weights and biases of a network. The learning rule is applied to train the network to perform a particular task. As the inputs are applied to the network, the network outputs are compared to the targets. The used algorithm performs clustering operations. They categorize the input patterns into a finite number of classes.

What we Expect from our New Method ?

Since ANNs learn from the data, the data must be valid for the results to be meaningful. Sometimes the desired response is unknown. Our model implements and uses the basic building blocks of neural computation, such as multi-layer perceptrons, Kohonen, Jordan and Elman networks, radial basis function, also called probabilistic networks, principal component analysis networks, self-organizing feature map networks and time-lagged recurrent networks. In our work we usually use NeuroSolutions, a *Neuro Dimension Inc. product.* A successful neural network simulation requires the specification of many parameters.

The performance is highly dependent on the choice of these parameters. A productive way to assess the adequacy of the chosen parameters is to observe the signals that flow inside the network. One can observe signals flowing in the network, weights changing, errors being propagated, and most importantly the cost, all while the network is working. This means that we do not need to wait until the end of training to find out that the learning rate was set too high. All probes within NeuroSolutions belong to one of two categories — static probes and temporal probes. The big difference is that the first kind access instantaneous data, while the second access the data over a window in time. By generating the code for the network, we can compile this code on a high-end workstation and train the network there. The resulting weights can then be saved to a file and imported back into your breadboard within NeuroSolutions.

CONCLUSIONS

Concentrated efforts are now focused in our labs in order to find original and modern solutions to obtain soft and hard nanostructured magnetic materials, and also to understand the intrinsic mechanisms governing their specific magnetic behavior. Modelling has now a very important position in the development and improvement of new materials for applications. Our knowledge of nanomaterials such as quantum dots, nanostructures, molecular assembly, is still at a very begining stage. Modelling and simulation techniques affect all stages in the development and improvement of new materials, from the initial formation of concepts to synthesis and characterization of properties.

We think new modelling strategies for complex material systems are needed. A new IT strategy ANN based for the nanomagnetic materials was given. Neural networks have been applied successfully in the identification and clasiffication of some nanomagnetic characteristics from large amounts of data. Neural networks was used for both regression and classification. The universal approximation capabilities of the multilayer perceptron make it a useful choice for modelling nonlinear systems and for implementing general-purpose controllers and magnetic characteristics extractor from wide data amount. Our model will permit the search of new materials with enhanced Hall effect - AHE and other magnetoresistive properties. The research is intended to set the foundation for more long-term work on on-line learning and construction techniques for the magnetic nanomaterials modelling.

7

Biomedical Aspects of Magnetic Nanoparticles

Though all the components of the body are either dia-, para, superpara- or ferromagnetic, the magnetic fields required which are likely to produce obvious effect on the body are very large. Even red blood cells with micrograms each of the iron protein haemoglobin are relatively unreactive to large fields or steep field gradients. However, there is sufficient iron present for MRI to be possible without adding ironrich or other contrast enhancing reagents. Indeed many cell types contain magnetite or other iron oxide nanoparticles—it is not clear whether these are generated by some standard biochemical process or are adventitious particles acquired from the environmnt.

The other natural iron-containing compounds in the body are haemosiderin, ferritin, transferrin and the cytochromes. The question of whether there are hazards or effects from magnetic fields of the magnitudes normally encountered in the environment een in the 'built' environment acting on these molecules is unlikely though unresolved. But the electromagnetic fields produced by mobile phones may perhaps be large enough to have pathological effects (see Cranfield *et al.* 2003 for a suggestion as to mechnism).

However the most likely exposures to large fields or field gradients arises from therapy or diagnosis. MRI produces images of both hard and soft organs and this is useful in diagnosis and in following the course of therapy. MRI contrast reagents re often used and there are a number of potential compounds for infusion or injection. Basically there are three ways in which a contrast reagent may act. It may not enter or attach to cells and simply act as an indicator of fluid-filled space eg the kneecapsule. It may attach to the outside of cells or lastly it may be taken into the cells. Closely related to these biological differences are the routes by which the agent reaches the target area. A wider and wider range of effective contrast agents contine to be developed and perhaps noteworthy amongst these are dendrimer based agents to help image the lymph system (Kobayashi *et al.* 2003), putrescine-Gd-amyloid beta peptides for Alzheimer plaque detection (Poduslo *et al.* 2003).

There are advantages in usng strongly magnetic nanoparticle contrast reagents, at least in theory, because there is a possibility that the particles can be localised in the desired region by applying local magnetic field gradients. There is also the possibility of producing localied hyperthermia when an organ loaded with nanoparticles is exposed to electromagnetic radiation. There are, however, potential if not real hazards with the use of superparamagnetic nanoparticles which are discussed below.

The localizability of the partices by applying external small fields is of apparent value in other types of therapy, for instance drug delivery or hyperthermia. It has been usual to suppose that the route or delivery would be by injection either through a vascular route or by localised njection into a tissue. After injection the particles would either remain un-attached to cells and drift around the body fluids. If attachment was reversible particles might move from cell to cell. The next phase would be the endocytosis of the particles nto the cells.

If the cells remain alive the particles are very unlikely to re-enter the extracellular space. The final localisation of the particles would be intracellular in many cases. Though it is clear from in vitro and some in vivo experiments that ells can survive the presence of such particles in moderate loadings for moderate periods of time, little is known about this. However, data from 'epidemological' sources suggests that carbon nanoparticles particles in the form of soot may present a risk t high exposures. On the otherhand 'natural' garden fires produce carbon nanotubes. Post-mortem samples should be taken from the lungs of people to discover if there is nanoparticle loading in the cells and if so whether it is associated with disease. Magetite or maghemite particles may present an additional risk because the pH in the endo-lysosomes (see Fig 1a) is low. This final destination is so acidic that ferric ions may be formed which will damage many biochemical processes. Silica coated nanopartices may be resistant to this effect.

Drug delivery is another perceived use of superparamagnetic particles. The drug would be attached to a surface coating on the particles such as dextran, PVA, polyglycols etc by fairly standard derivatisation chemistry.Again the application of external magnetic fields would be used to 'capture' the particles in the desired site. Besides the localizability of the particles the immobilisation of a signal molecule on the particles may protect it to an extent from degradatin and the long term binding to a receptor may lead to that receptor being activated for long periods. Thus use of nanoparticles may prove to be a very efficient method of delivering a drug to a cell. There are two possible events, the one usually reportedis the endocytosis of the particles after preliminary capture at the surface of the cell.

The other, discovered only recently is that if certain coating proteins are attached to the particle endocytosis is prevented. This requires that the proteins be boud by surface receptors. This is not in itself a disadvantage in many situations as a very large number of drug receptors are located at the cell surface and since one of them can be chosen to bind the particle to a surface receptor irreversibly and anothe can be attached to the first to be the active drug which will attach to a nearby second receptor. The advantage of this system is that endocytosis does not occur (see Fig 1b) and there is at least the possibility that use of a competing soluble ligand ma allow removal of the signal.

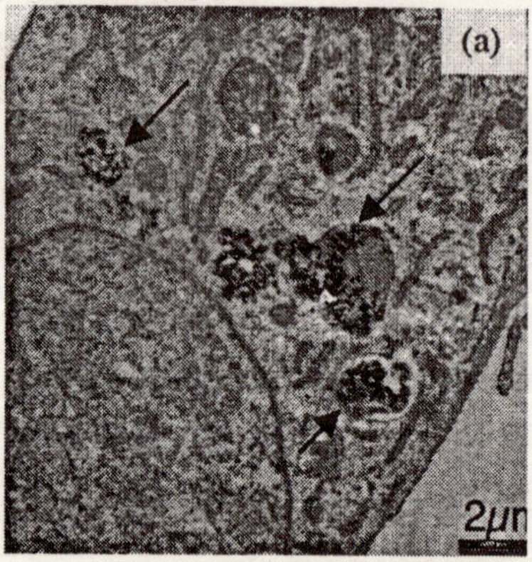

Fig 1a Transmission electron micrograph section (TEM) of a cell (h-tert human fibroblast) that has endocytosed uncoated magnetite nanoparticles, Note their location in the phagolyzosome, arrows indicate phagolysosomes.

So in terms of drug delivery magnetic nanoparticles offer the possibility of use of external magnetic fields to obtain better localisation than could be achieved with non-magnetic particles. But since magnetic nanoparticles re less easily destroyed or inactivated by cells than many non-magentic ones there is the disadvantage that persistent particles may cause later cell damage and death. The same considerations apply to situations where magnetic nanoparticles are being usedfor generating hyperthermia by the application of external fields.

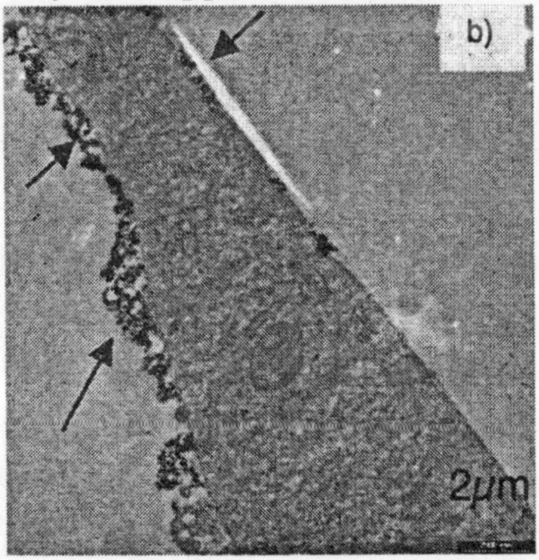

Fig 1b TEM of a cell (h-tert human fibroblast) exposed to lactoferrin labelled nanoparticles after 24 hours. The particles are on the cell surface or close outside it) see arrows) and none have been endocytosed. Claims have been made in the literature that transferrin-labelled (a related protein) particles are endocytosed but the evidence for this is often based on light microscopy which does not have sufficient resolution to locate the particles precisely. Arrows mark cell surface accumulations.

Cartmell *et al.* (2003) have suggested that mechanical stimulation could be applied to cells by using an external magnetic field to apply a twisting motion to the particles. Since mechanicl stimulation is effective in activating cells this may be a future area for therapeutic developments.

One interesting if rather worrying possibility is that exposure of cells to nanofeatures or nanoparticles can cause large, if perhaps only temporary, canges in gene expression. This would not necessarily be linked solely to the use of magnetic nanoparticles because the work of Dalby *et al.* (2003) shows that nanofeatures on polymer surfaces can cause change in gene expression. Since excess particles may well be adsorbed onto the surfaces of cells and intercellular material it is appropriate to consider how these adsorbed particles might affect cell behaviour. This has hardly been examined but some recent work (Berry and Curtis 2003) suggests that they ma act much as nanofeatured surfaces do.

MESOSCOPIC AND NANOSCOPIC STRUCTURES

*A*ll the magnets currently used are based on metallic or ionic lattices. Since about two decades chemists have started a thorouh project of using molecular chemistry techniques to develop new classes of magnets based on molecules rather than on metals or oxides.

The idea behind this is the challenge of creating new classes of materials from which new exciting properties may be exected. In a sense this research is the continuation of that which was successfully developed when it was discovered that organic compounds can behave as conductors and superconductors like the classic inorganic materials.

In a few years it has been discovred that purely organic magnets are indeed possible, although the critical temperatures are still very low. The most promising results have been obtained using sulfur-nitrogen based organic radicals which behave as weak ferromagnets below 35 K. On theother hand, using a mixture of transition metal ions and organic radicals it has been possible to obtain a room temperature ferrimagnet, and similar results have been obtained using derivatives of the old Prussian Blue compounds.

Beyond the chemial challenge of assembling new structures using moderately stable building blocks like organic radicals, the factors suggesting that molecular magnetism can indeed provide new interesting classes of materials are:

- The possibility of fine tuning the prperties of the materials by using flexible molecular techniques
- The possibility of building a la carte magnetic molecules of increasing size which behave as molecular nanomagnets
- The possibility of obtaining multifunctional materials

An example o the last possibility has been recently reported by Coronado *et al.* They used a hybrid approach, assembling together inorganic building blocks like transition metal oxalato derivatives and organic radicals like the famous BEDT-TTF, well known to form onducting and superconducting materials. They obtained by electrocrystallisation a compound of general formula $[BEDT\text{-}TTF]_3[MnCr(oxalato)_3]$. The structure comprises honeycomb inorganic layers of $[MnCr(oxalato)_3]$- separated by stacks of the organic radicals.

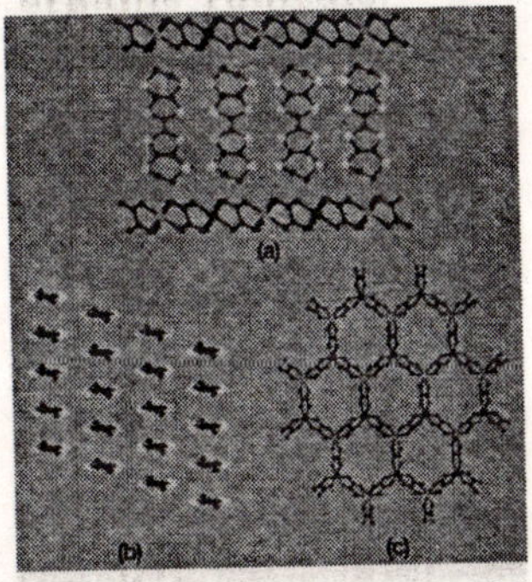

Fig 1 The structure of [BEDT-TTF]3$[MnCr(oxalato)_3]$. a) overview of the lattice; b) the organic layers; c) the inorganic layers.

The average charge on the BEDT-TTF molecules is +0.34. The inorganic layer is insulating while the organic moiety is a conductor. The magnetic coupling is fairly strong within the inorganic layers and this, coupled to the fact that eak interactions are operative between the layers determined a transition to bulk ferromagnetic behaviour below 5.5 K. Therefore below the critical temperature the material behaves as a ferromagnetic conductor. Although systems like this are well known strting from iron itself, in the molecular derivative the magnetic electrons are different from the conducting electrons, thus offering the possibility of observing new phenomena.

Another important feature of molecular magnets is that they are in general isulators, therefore they are much more transparent to UV-visible light than classic magnets. Therefore it is possible to use light to induce magnetic transitions.

This approach has been used by the groups of Verdaguer and Hashimoto for instance.Prusian blue derivatives are complex cyanides of general formula ABC(CN). When B= Fe^{2+} and C= Co^{3+} the compound is diamagnetic because both ions are in their low spin, non-magnetic state. By illuminating with red light however it is possible to induce an elecron transfer in which Fe^{2+} is changed to low spin Fe^{3+}, with one unpaired electron, and Co^{3+} to high spin Co^{2+} with three unpaired electrons:

$$Fe^{2+}\text{-C-N-}Co^{3+} \text{ - } Fe^{3+}\text{-C-N-}Co^{2+}$$

Th material orders as a bulk ferrimagnet below 50 K. If the irradiation is performed below this temperature we observe a transition to bulk magnetic order induced by light. Therefore these materials can be considered as magnetic switches operated by light. t is also possible to perform the opposite transition by irradiating the Fe^{3+}-Co^{2+} pairs with blue light: the electron is back transferred from cobalt to iron and the system reverts to thediamagnetic state.

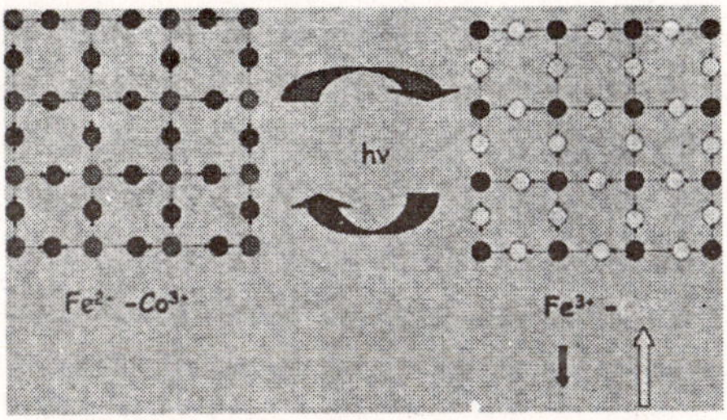

Fig 2 Scheme of the light induced interconversion of Prussian blue derivatives.

The most exciting developments of molecular manetism of the last few years has been the discovery that some discrete molecules can behave at low temperature as tiny magnets. The archetypal molecule is:

$[Mn_{12}O_{16}(CH_3COO)_{16}(H_2O)_4].4H_2O.2CH_3COOH$, Mn12Ac for short.

It comprises an external ring of eight manganese(III) ions and a tetrahedron of manganese(IV) ions, bridged by oxide bridges, like in a manganite. Acetic acid molecules complete the environment of the metal ions, infact blocking the growth of the oxide particles. In other terms these clusters can be considered as small particles of manganese oxides, all identical to each other. In this particular case the particles crystallise in a tetragonal space group, and therefre they are also iso-oriented in the lattice. At low temperature the system is in the ground S= 10 state, with the eight S= 2 spins of manganese(III) up and the four S= 3/2 of the manganese(IV) ions down.

The clusters have a high Ising type magnetic anisoropy, due to the local anisotropies of the distorted manganese(III) ions. This means that the magnetisation is preferentially oriented parallel to the tetragonal axis of the cluster and to reorient itself it needs a high activation energy. At 2 K the relaation time of the magnetisation becomes so long that the individual molecules become permanently magnetised, like tiny magnets. They show magnetic hysteresis and in principle information can be stored in one molecule, thus reaching the highest possible leel of miniaturisation. But the excitement does not stop here, because Mn12Ac shows a rich range of quantum properties, and it has been proved to be an ideal test ground for theories of quantum effects in magnets. Clear evidence of tunnelling of the magnetsation has been achieved by observing stepped hysteresis. More controversial still is the possibility to observe quantum coherence.

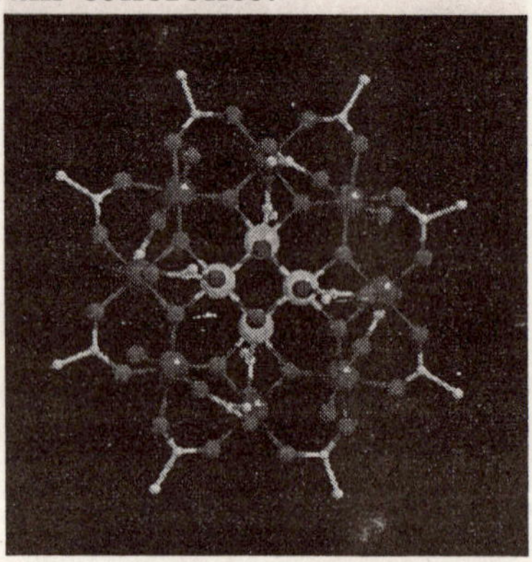

Fig 3 Sketch of the structure of Mn12Ac.

The advantage of magnetic molecules over other types of magnetic particles is that they are absolutely monodisperse, and tat in principle they can be diluted and organised using supramolecular chemistry techniques. Many efforts are currently made to derivatise the magnetic molecules to organise them on suitable supports, like gold or silicon, in order to be able to address tem individually. The dream of storing information in one molecule may become reality!

After Mn12 a few other types of single molecule magnets have been discovered and investigated. Among these a cluster containing eight iron(III) ions has been found to sow oscillations of the tunnel splitting in the presence of a an applied transverse magnetic field. This is the signature of the Berry phase in magnets which was first reported by Sessoli and Wernsdorfer in 1999.

Recently it has been discovered that smilar effects of slow relaxation can be observed in one-dimensional magnets. A compound containing Co^{2+} ions regularly alternating in space with organic radicals behaves as a one dimensional ferrimagnet. Below 20 K the relaxation time of the magnetisaion rapidly increases on decreasing temperature, with a barrier for the re-orientation of the magnetisation larger than 150 K. Also in this case magnetic hysteresis is observed which is associated with a high magnetic anisotropy of the easy-axis type. A smi quantitative explanation has been provided using a model proposed by Glauber in 1963. This opens an exciting perspective, that of allowing the storing of information in segments of a polymer, which therefore behaves like a magnetic nanowire.

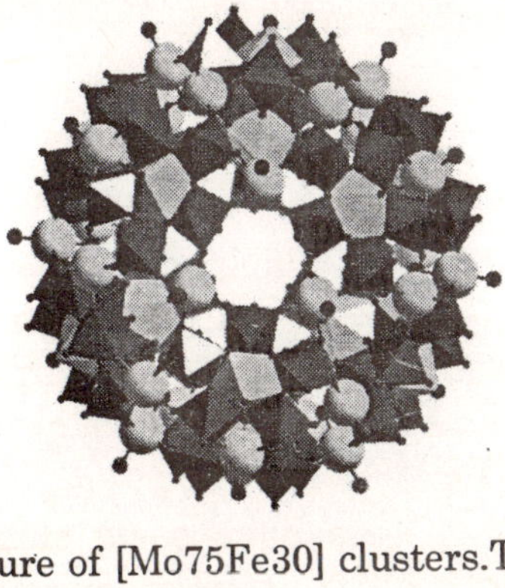

Fig 4 The structure of [Mo75Fe30] clusters.The yellow circles are iron(III) ions.

Finally Iwant to mention that magnetic molecules are also interesting in the case of antiferromagnetic behaviour. Perhaps the most significant example is that of a cluster containing 76 Mo^{6+} ions and 30 Fe^{3+}, which has the structure shown in Figure 4. The molbdenum ions are non-magnetic so the magnetic behaviour is associated with the interactions between the iron ions. They are antiferromagnetic, and highly frustrated due to the particular topology which corresponds to an icosidodecahedron. The compound has ot shown quantum effects down to the mK region, due to the large degeneracy of the low lying states associated with spin frustration.

Molecular magnets open many new possibilities to observe quantum and classic effects in mesoscopic matter. The field isjust at its beginning, but many developments can be anticipated in the next few years. Possible applications may range from quantum computing to new types of contrast agents for magnetic resonance imaging.

8

Ferromagnetic Semiconductor Heterostructures

INTRODUCTION: DILUTED MAGNETIC SEMICONDUCTORS

*A*s discussed in accompanying papers, today's research on spin electronics involves virtually all material famiies, the most mature being studies on magnetic metal multilayers, in which spin-dependent scattering and tunnelling are being successfully applied in reading heads of high density hard-discs and in magnetic random access memories (MRAM). However, in the cntext of spintronics particularly promising are ferromagnetic semiconductors since they combine complementary resources of ferromagnetic and semiconductor material systems.

One of the relevant questions is to what extend the powerful methods develped to manipulate the carrier concentration and spin polarisation in semiconductor quantum structures could serve to tailor the magnitude and orientation of magnetisation produced by spins localized on magnetic ions.

Another important issue concerns the eaboration of methods of injecting and transporting spin currents, which may ultimately lead to control over single spins in solid state environment.

Already early studies of Cr spinels as well as of rock-salt Eu- and Mn-based chalcogenides led to the obervation of a number of outstanding phenomena associated with the interplay between ferromagnetic cooperative phenomena and semiconducting properties. A strong spin-dependent (exchange) coupling between the band carriers and spins localized on the magetic atoms accounts for outstanding properties of these materials.

This coupling gives rise to strong indirect exchange interactions between the localized moments as well as to giant spin-splittings of the electronic states, which is proportional to magneisation of the spins. The discovery of ferromagnetism in Mn-based zinc-blende III-V compounds, such as (Ga,Mn)As, followed by the prediction and observation of ferromagnetism in p-type II-VI materials, such as (Zn,Mn)Te:N, allows one to explre the physics of previously not available combinations of quantum structures and ferromagnetism in semiconductors. This aspect of ferromagnetic III-V and II diluted magnetic semiconductors (DMS) will be outlined here together with indications of models aming at explaining the nature of ferromagnetism in these materials.

CARRIER-CONTROLLED FERROMAGNETISM

It is well established that Mn is divalent in II-VI compounds and assumes the high spin d^5 configuration characterized by S = 5/2. Here, Mn ions neitherintroduce nor bind carriers, but give rise to the presence of localized spins. For low carrier densities, II-VI DMS are paramagnetic but neighbour Mn-Mn pairs are antiferromagnetically blocked owing to short-range super-exchange interactions. However, thi antiferromagnetic coupling can be overcompensated by ferromagnetic interactions mediated by band holes. In III-V compounds,

Mn atoms when substituting trivalent metals supply both localized spins and holes, so that extrinsic co-doping is not necessry to generate the carrier-mediated spin-spin interaction.

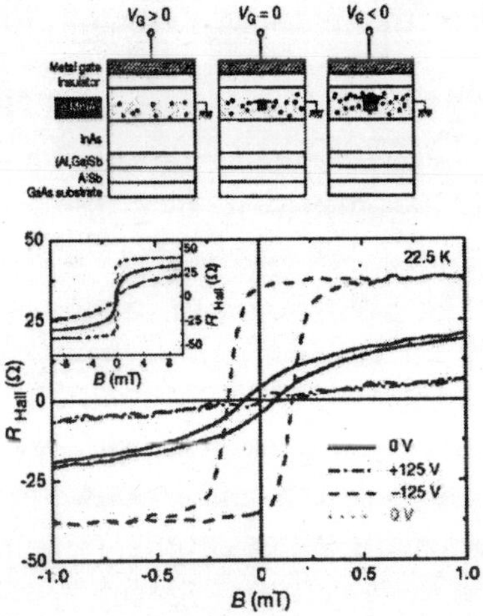

Fig 1 Magnetisation hysteresis loops determined by measurements of anomalous Hall effect at constant temperature of 22.5 K for various gate voltages in field-effect transistor structure with (In,Mn)As channel. (after Ohno *et al.*).

The role of band carriers in promoting ferromagnetic ordering between localized spins was already noted by Zener in the '50s in the context of magnetic metals. This ordering can be viewe as driven by the lowering of the carriers energy associated with their redistribution between spin subbands, split by the exchange coupling to the localised spins. A more detail quantum treatment indicates, however, that the sign of the interaction betwen localised spin oscillates with their distance according to the celebrated Ruderman-Kittel-Kasuya-Yosida (RKKY) model. Because of the large density of states and spin-dependent hybridisation between anion p and magnetic d states, the carrier-mediated spi-spin interaction is particularly strong in the presence of the holes in tetrahedrally coordinated DMS. Interestingly, this carrier-mediated Zener/RKKY ferromagnetism is enhanced by exchange interactions within the carrier liquid, so that the Stoner mechaism contributes to the magnitude of the Curie temperature in these systems.

MANIPULATIONS OF MAGNETISATION

Since magnetic properties are controlled by band holes, an appealing possibility is to influence the magnetic ordering isothermally, by light or bythe electric field, which affect the carrier concentration in semiconductor structures. Such tuning capabilities of the materials systems in question were put into the evidence in (In,Mn)As/(Al,Ga)Sb and modulation doped p-(Cd,Mn)Te/(Cd,Mg,Zn)Te heterostructures. Actually, these findings can be quantitatively interpreted by considering the effect of the electric field or illumination on the hole density under stationary conditions and, therefore, on the Curie temprature in the relevant magnetic layers. Interestingly, according to experimental findings and theoretical modelling, photocarriers generated by above barrier illumination destroy ferromagnetic order in the magnetic quantum well residing in an undoped (intinsic) region of a p-i-p structure but they enhance the magnitude of spontaneous magnetisation in the case of a p-i-n diode.

For the valence band states, whose periodic part of the Bloch functions contain spin components ixed up by the spin-orbit interaction, the exchange splitting does not depend only on the product of the p-d exchange integral and the Mn magnetisation but also on the magnitude and direction of the hole wave-vector, confinement, and strain. In particular both experimental and theoretical results demonstrate that the orientation of the easy axis in respect to the film plane depends on whether the epitaxial strain is compressive or tensile. Hence, magnetic anisotropy and thus easy axis direction and doain structure, can be manipulated by appropriate layout of the layer sequence, as epitaxial growth of DMS films in question is usually pseudomorphic. Furthermore, magnetic anisotropy at given strain is predicted to vary with the degree of the occupaton of particular hole subbands. This, in turn, is determined by the ratio of the valence band exchange splitting to the Fermi energy, and thus, by the magnitude of spontaneous magnetisation, which depends on temperature. Such a temperature-induced switchig of the easy axis direction has recently been detected in samples with appropriately low hole densities. A related sensitivity of the easy axis direction to the carrier concentration constitutes a novel method enabling control of the magnetisation drection locally by a system of electrostatic gates.

SPIN INJECTION

A number of groups is involved in the development of devices capable of injecting spins into a non-magnetic semiconductor. Obviously, owing to a high degree of spin polarisation and resitance matching, ferromagnetic semiconductors constitute a natural material of choice here. Typically, a p-i-n light emitting diode structure is employed, in which the p-type spin injecting electrode is made of a ferromagnetic semiconductor. In this particular experiment, the degree of circular polarisation is examined for light emitted in the growth direction. In this Faraday configuration, simple selecton rules are obeyed for radiative recombination between the electron and heavy hole ground state subbands. Since the easy axis is in plane, a field of a few kOe is necessary to align the magnetisation and thus to produce a sizable degree of light polarisaion. Importantly, the injection of spin polarised electrons, using Zener or Esaki tunnelling from p-type (Ga,Mn)As electrode into n-type GaAs, was also realized.

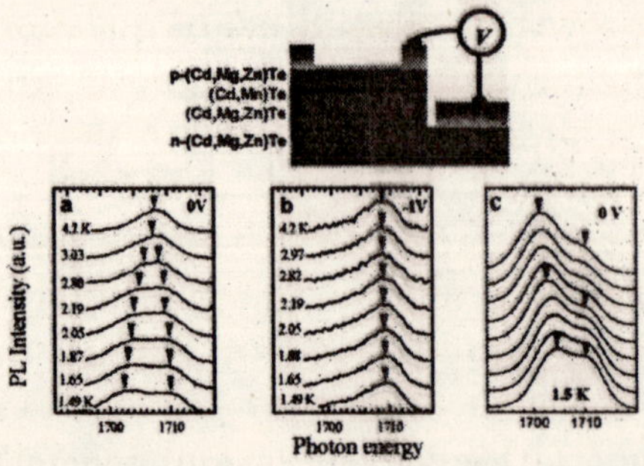

Fig 2 Effect of temperature (a), bias voltage (b), and illumination (c) on photoluminescence of structure consisting of modulation doped p-(Cd,Mn)Te quantum well and n-type barrier. Zero-field line splitting (marked by red arrows) and shift on lowering temperature witness the appearance of a ferromagnetic ordering at low temperatures (a), which does not show up if the quantum well is depleted from the holes by reverse bias of the pi- n diode (b).The low-temperature splitting is enhanced by additional illumination by white light which increases the hole concentration in the quantum well (after Boukari *et al.*).

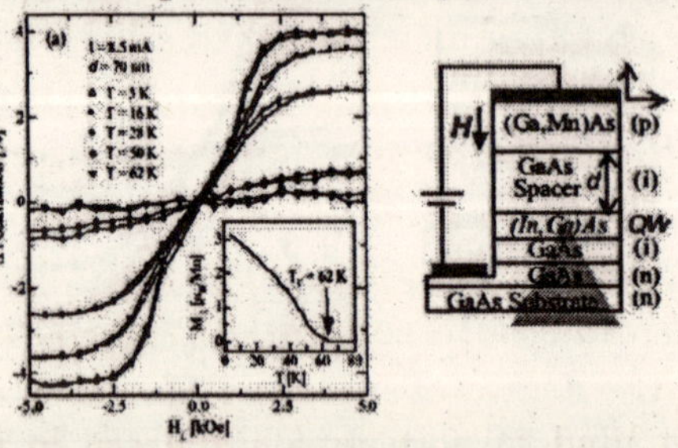

Fig 3 Degree of circular polarisation of light emitted by a (In,Ga)As quantum well located in a p-i-n diode biased in the forward direction and containing (Ga,Mn)As as the p-type electrode. An external magnetic field is applied along the hard axis, that is perpendicularly to the interface plane.The degree of circular polarisation and the (Ga,Mn)As magnetisation depend similarly on the magnetic field and temperature, which together with the lack of photoluminescence polarisation for excitation

by linearly polarised light, point to the existence of the hole spin injection from ferromagnetic (Ga,Mn)As to non-magnetic (In,Ga)As via a non-magnetic GaAs (afterYoung *et al.*).

Quantum Hall Ferromagnet

The above results demonstrate rather convincingly the decisive roleof the valence band holes in setting on the ferromagnetic ordering in these systems, a conclusion corroborated by studying the Curie temperature TC as a function of compensation in epitaxial films of (Ga,Mn)As and as a function of the acceptor concentratin in p-(Zn,Mn)Te. At the same time, in agreement with theoretical predictions, no ferromagnetic ordering was detected above 1 K in heavily doped n-type (Zn,Mn)O. However, according to experimental results presented for an appropriate combiation of an external magnetic field and gate voltage that controls electron concentration, an Ising ferromagnetic ground state with TC up to 2 K can be observed in high-quality modulation-doped n-(Cd,Mn)Te/(Cd,Mg)Te heterostructures.

The quantum Hallferromagnetism shows up when two partly filled Landau levels with opposite spin directions are brought into a coincidence. This Stoner-like instability of the carrier liquid appears due to the enhancement of the density of states in two-dimensional system in quantising magnetic fields. Incidentally, Landau level crossings occur in such structures due to the enlargement of electron spin-splitting by the s-d exchange interaction with the Mn spins.

SUMMARY AND OUTLOOK

Extensive studies of ferromagnetic semionductor heterostructures have lead to the observation of a number of spin phenomena. Some of them, such as spin injection as well as giant and tunnelling magnetoresistance, are familiar from parallel studies of magnetic multilayers. Other effects, like iothermal driving of the system between ferromagnetic and paramagnetic phases by the electric field or by the photon flux in the direction that can be selected by an appropriate design of the structure, appear unique to ferromagnetic semiconductors.

These triking effects offer new tools for patterning magnetic nanostructures as well as for information writing and processing, beyond the heating effects of light exploited in the existing magneto-optical memories. Obviously, however, while the potential of feromagnetic semiconductors can already be exploited for the development of quantum information hardware, their practical applications in classical information systems have to be preceded by progress in the synthesis of a functional material with TC surpassng comfortably the room temperature.

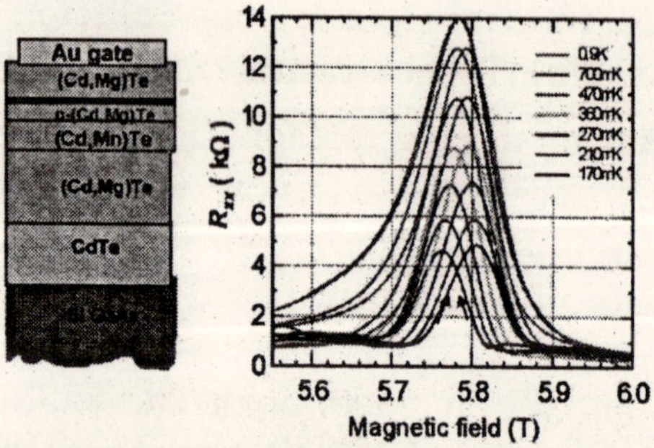

Fig 4 Quantum Hall resistances in modulation doped p-(Cd,Mn)Te quantum well containing a metal gate for changing electron concentration. Resistance spike at 5.8 T occurs at the crossing of ground and excited Landau levels with opposite spin orientations (0↑ and 1↓).The presence of the spike and its hysteretic behaviour signalises ferromagnetic ordering of the electrons residing in partly filled overlapping Landau levels (after Jaroszynski *et al.*).

Following the theoretical suggestion, a number of oxide and nitride-based DMS, containing Mn or other magnetic elements, has been obtained, which indeed show indications of ferromagnetism at room temperature. owever, a further progress in this field requires development of growth and characterization methods that enable a better control over solubility limits, self-compensations, phase segregations, and precipitations of other compounds.

On the theoretical sde, it appears that double- and super-exchange, rather than the Zener/RKKY mechanism, may account for ferromagnetism is systems containing magnetic elements other than Mn. Since ferromagnetic DMS combine intricate properties of charge-transfer insulators nd strongly correlated disordered metals with the physics of defect and band states in heavily doped semiconductors, despite important advances in theoretical understanding of these systems, their description from first principles may take some time. Withut any doubt, the search for functional ferromagnetic semiconductor nanostructures and their theoretical modelling has evolved into an important branch of today's materials science and condensed matter physics.

MAGNETIC MATERIALS: FROM THE SEARCH OF NEW PHASES TO NANOSCALE ENGINEERING

From Soft to Hard Magnetic Materials

There is a category of materials, of which the magnetisation is used as a source of a magnetic field or of a magnetic flux. All these materials are Fe, Co or Ni based (trasition metal elements, M) and their spontaneous magnetisation, M_s, have similar typical values. Yet, they may exhibit extremely different applicative properties. This is due to differences in the strength of their coercivity, i.e. their ability to resist an applied magnetic field, $\mu_0 H_{app}$, which tends to change the orientation of the magnetisation.

From extremely soft to extremely hard materials, the coercive field, μ_0H_c, at which the magnetisation finally follows the applied field, increases by more than 6orders of magnitude (see Figure 1). Soft materials are used in magnetic circuits, transformers or sensors; hard materials are used to make motors, actuators or generators; materials with medium coercivity are used as storage media. Coercivity depends on material intrinsic magnetic properties (temperature below which ferromagnetism is stable, magnetisation and anisotropy energy (see below)). It depends also on extrinsic properties (grain size, grain orientation, grain separation, etc...), which define thematerial microstructure.

During the last 50 years, an impressive development has occurred in the field of magnetic materials, stimulated by the better understanding of the characteristic behaviours of the different types of magnetic elements. More recenty, the discovery of behaviours which are specific to nanomaterials, has open a new period. At the nanoscale, the distinction between intrinsic properties, characteristic of a given magneti phase, and extrinsic properties, related to the detail of the micostructure, disappears.

Interactions Involved and Fundamental Processes

The typical equilibrium magnetization configuration of a ferromagnetic material is made of micron-size domains. Within a given domain, the moments are parallel as required by exchang interactions and they are aligned along a particular direction, the easy magnetisation direction, as required by magnetocrystalline anisotropy. From one domain to the next, the magnetisation direction alternates. Globally, the magnetisation is zero; exchnge and anisotropy energies are minimized, as they are in the single domain state, except within small regions, the domain walls, which constitute the transition regions between two magnetic domains. Additionally, magnetic domain formation allows the weakr long-range dipolar interactions to be strongly reduced with respect to their value in the single-domain state.

Under an applied field, a bulk magnetisation tends to develop to minimize Zeeman energy. In an ideally homogeneous material, the domain wall nergy does not depend on the wall position. The magnetisation increases by free domain wall motion, such that the domains in which the magnetisation is along the applied field grow at the expenses of the others. Such materials are magnetically soft. The induced magnetisation is in turn the sourcc of a magnetic field or of a magnetic flux. When the applied field is time dependent, the magnetisation is time-dependent as well. A voltage results, which constitutes the signal exploited in transformer or sensors.

When the applied field is further increased, one reaches ultimately a state in which only one domain remains. The magnetisation is saturated. Assuming that a field is applied again, but of reversed sign, magnetisation reversal requires thata domain of reversed magnetisation is nucleated. The value of the nucleation field is closely linked to that of the anisotropy energy. This can be understood by considering that a moment that participates in the magnetisation processes necessarily assumesan intermediate transitory position in which it is perpendicular to the easy magnetisation direction.

In this unfavoured orientation, the magnetocrystalline anisotropy energy is at a maximum. Permanent magnets are based on high anisotropy materials. A potential energy can be stored in a field, which is antiparallel to the magnetisation. Motors or actuators transform this potential energy into mechanical energy.

The two states of saturated magnetisation, up and down, may constitute the two elemetary bits of information, 0 and 1. Materials used for information storage have medium coercivity: the magnetisation-saturated states must be stable enough for the information not to be lost; at the same time, the writing process must be realised in a not oo high magnetic field. Exchange decoupling between the constituent grains is a specific feature of these materials. It allows each bit of information to be written independently of the others.

Which Materials?

Ferromagnetism above room temperture is a unique property of the 3d transition metals (M= Fe, Co or Ni), their alloys or compounds. It results from large exchange interactions existing between the magnetic 3d electrons. In general, alloys and metallic compounds have higher magnetisaion than insulating compounds. At room temperature, pure Fe metal has the largest magnetisation of all known elements ($\mu_0 M_s = 2.2\ T$).

Materials where 3d magnetism only is involved tend to have low magnetic anisotropy and often show soft magnetic behavior. In the best of these materials, the permeability, which represents the ratio between the induced magnetisation and $\mu_0 H_{app}$, may be as high as 10^5. Most of these materials have been known for many years. This is in particular the case of the Fe-Si alloysin which 3-5 % Si is substituted for Fe. The magnetisation is approximately 10 % less than that of pure Fe metal.

Other properties (mechanical properties, resistivity, corrosion resistance, etc...) are decisively improved with respect to those of the pue metal. Fe-Si sheets constitute the core of all transformers used for energy transportation. Soft magnetic materials are mostly used under AC exciting fields. At high frequency, eddy-current losses develop in metallic systems. Insulating ferrites are use in electronics (AC-DC converters, inductances, etc...) at frequencies from typically 1 kHz to 1 GHz. At even higher frequencies, rare-earth garnets are used in non-reciprocal devices, for telecommunication applications.

The hard ferrites are the first manets that exploited the link existing between coercivity and magnetocrystalline anisotropy. These materials are used in the fabrication of small motors used in the automotive industry and for home appliances. In the 1960's, extremely large anisotropie were discovered in the rare-earth metals (R = rare-earth), the second series of magnetic elements in the periodic table. The R elements are heavy elements, and the magnetic 4 f electrons are characterised by large spin-orbit coupling. By this mechanism te moment orientation is tightly linked to the crystal structure, i.e. the magnetocrystalline anisotropy is large.

The interatomic 4f-4f exchange interactions are weak and magnetism in the R metals is only stable at low temperature. In the R-M compounds, srong 4f-3d interactions exist, mediated by the rare-earth 5d electrons. Rare-earth magnetism is preserved above room temperature and, with it, compounds with very large anisotropies may be obtained. Magnets based on the SmCo5 phase, with typical compositin $Sm(Fe,Co,Cu,Zr)_{7-8}$, are of particular interest for high-temperature applications. The fabrication of magnets based on the $Nd_2Fe_{14}B$ phase constituted a decisive breakthrough in the development of rare-earth based magnets. These magnets associate utstanding magnetic properties ($\mu_0 M_s = 1.2\text{-}1.4\ T$, $\mu_0 H_c$ *1-2 T*) with much lower cost than the $SmCo_5$ magnets. They are used in particular in compact high performance motors, such as the voice coil motors of all hard-disk drives.

Pd or Pt, alloyed with M metls, may also be the source of high anisotropy. In these heavy elements, the magnetic electrons are the 4d or the 5d ones. The equiatomic $L1_0$ Fe-Pt phase has extremely large anisotropy. High performance materials have been prepared in thin film forms. Magnts inserted in microsystems could find application, despite the high cost of Pt.

The coercive field required in materials for magnetic storage is approximately an order of magnitude smaller than the coercive field of hard magnets, $\mu_0 H_c \approx 0.05 - 0.4$ T. Such coercive field values can be obtained in materials, which show shape anisotropy: the moments tend to align along the long dimension of elongated particles. The γ-Fe_2O_3 particles, which constitute the most common material used in magnetic tapes, exploitthis property. More recently, elongated Fe metallic particles, which have higher magnetisation and coercivity, have been developed for the same purpose.

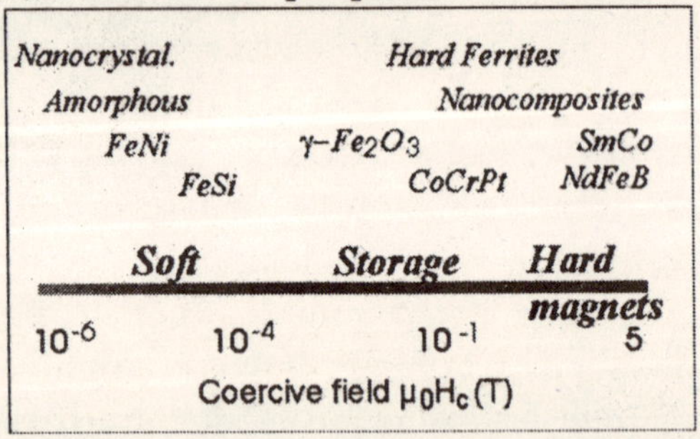

Fig 1 Magnetic materials according to their coercivity.

The materials used in Hard Disk Drives (HDD) are Co-based alloys, which contain additive elements suc as Cr, Pt and B. The magnetic grains are coated with a non-magnetic Cr-rich layer, which ensures exchange-decoupling. Their coercivity is related to magnetocrystalline anisotropy. The grains being very small, a significant fraction of the atoms are srface atoms.

These are known to potentially show larger anisotropy than bulk atoms.

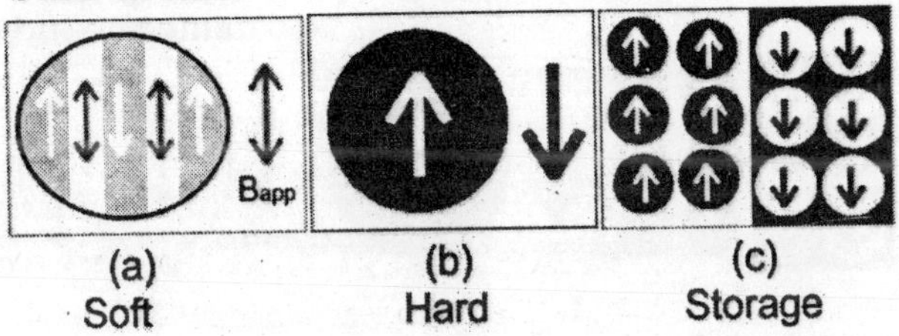

Fig 2 Magnetic materials of different kinds (a) Soft materials exploit the magnetisation variation with the applied field (b) Hard magnets exploit their ability to resist to a reversed magnetic field (c) in materials for information storage, the magnetisation is stable to keep the information stored and it can be reversed to change this information.

Amorphous alloys, which associate R elements (Gd, Tb) to M elements (Fe, Co) have been developed for perpendicular magneto-optical recording. The R and M moments are oupled antiparallel and the chemical composition is such that, at room temperature, the global magnetisation is almost zero. This allows the magnetisation to lie perpendicular to the substrate surface. To first order, the rare-earth elements do not contriute to the magneto-optical reading signal, which is approximately proportional to the transition metal magnetisation.

To the Nanoscale

The understanding of the magnetism of matter is such that it appears unlikely that new alloys and compounds with signiicantly improved intrinsic properties may be discovered. Tailoring the magnetic properties at the nanoscale offers much more perspective. At this scale, new properties emerge because the dimensions are below the characteristic length-scales of magnetic ineractions or magnetic phenomena.

Let us consider an assembly of exchange-coupled randomly-oriented nanocrystals. At small scale, the dipolar interactions need not to be considered. Exchange favours parallel coupling of the moments over the whole sampe.

Anisotropy favours moment alignment along the local easy magnetisation direction of each nanocrystal. One expects in general that exchange energy is much higher than anisotropy energy. However, the nanocrystals are coupled through their surface and thu the exchange energy intervenes as a surface term, whereas anisotropy energy is a volume term as usual.

Due to the interplay between these 2 terms, a certain correlation volume must exist over which the moments can be considered as fully aligned. This volme may be expressed as $V\ A^6/K^6v^3$, where $A\ (10^{-12}\ J/m)$ is the exchange constant between nanocrystals, K is the nanocrystal anisotropy constant and v is the nanocrystal volume. The number of crystallite in V is not infinite and a resultant anisotrop energy exists at the scale of V, determined by statistics. This energy, normalised to one nanocrystal, is expected to be proportional to $1/N$, where N is the number of nanocrystals within the correlation volume.

Fe-based ultra-soft nanomaterials (such asFeSiBNbCu or FeZrB) are formed of α-Fe nanocrytals (10^{-15} nm in diameter) embedded into an amorphous magnetic matrix (Figure 3). For these systems, $K\ 10^{-4}\ J/m^3$, the correlation volume thus derived is of macroscopic size. The moments being aligned ovr macroscopic dimensions, the global magnetic anisotropy is averaged over a very large number of grains and it is vanishingly small. This explains the ultra-soft behaviour obtained. These nanomaterials are used to make transformers for electronic applicatons.

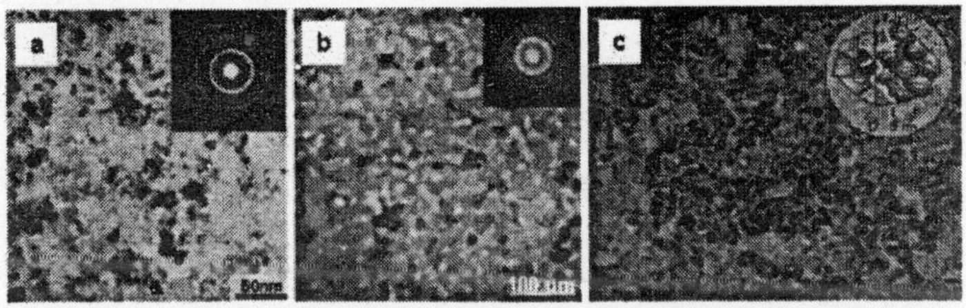

Fig 3 Three different types of nanostructured magnetic materials (a) FeSiBNbCu ultra-soft magnetic materials (b) NdFeB hard nanocomposites (c) CoCrPtB magnetic films for storage. In (a), the soft magnetic properties directly result for exchange coupling between the grains. In (b), exchange coupling between soft and hard grains is at the origin of the global coercivity obtained In (c), exchange decoupling between grains is characteristic of materials for magnetic recording so that the magnetisation of any given region may be reversed,without influencing others.

In R-M intermetallics, the anisotropy constant K $10^{-7}J/m^3$. The correlation volume deduced is of the order of 10^3 nm^3, i.e. of the order of or smaller than the nanocrystal size. To first approximation, the moments are aligned along the local easy agnetisation direction. In this configuration, the anisotropy is not averaged out. This is in agreement with the existence of large coercivity in NdFeB nanostructured alloys. In order to minimize exchange, the moments close to the interface between tw nanocrystals, tend to orientate along some intermediate direction between the 2 local easy magnetisation directions. It results that the magnetisation is higher than the value $M_s/2$ expected for perfect alignment along the individual easy magnetisation diections of an assembly of uniaxial randomly-oriented nanocrystals. This is the phenomenon of remanence-enhancement.

Hard nanocomposites include soft and hard exchange-coupled nanocrystals, in which the soft nanocrystals show coercivity (Figure 3). To explain this, let us assume that the applied field is weaker than the hard nanocrystal coercive field, so that the corresponding magnetization is frozen. Due to exchange coupling at the interface, soft phase magnetisation reversal requires the formtion of a moment configuration within the soft phase, which is very reminiscent of a domain wall.

However, this wall is constrained to the size of the soft nanocrystals. It is much narrower than it would be in the bulk and its energy is 1-2 orders of magntude higher. Thus, the soft phase resists reversal. The most common of these materials associate hard $Nd_2Fe_{14}B$ nanocrystals and soft α-Fe or Fe_3B ones; $\mu_0 H_c$ values up 0.7 T are obtained. When these materials were discovered, it was hoped that ne high performance materials could be produced by associating a coercive hard magnetic phase with large magnetisation soft phases. To date, the properties obtained do not merit large industrial development. The recent successful preparation of anisotropic anocomposites may open new perspectives.

Storage media consist of exchange-decoupled randomly-oriented grains. The transition region from one bit to the next must be sharp. This implies that each bit is made of a large number of grains, typically 100-10000. The recording density envisaged with magnetic tapes is of the order of 5 Gbits/in^2. The corresponding particle volume is around 10^5 nm^3. In modern HDD, the recording density approaches 100 Gbits/in^2. At this density, the bit size is 0.2 μm x 0.1 μmand the particle volume is around 100 nm^3. Thus, magnetic storage necessarily involves nanomaterials.

At such small particle sizes, the phenomenon of superparamagnetism must be considered. The total anisotropy energy of one nanoparticl, Kv, must be compared to the energy brought by thermal activation at the temperature T, of the order of $25k_BT$. Setting $Kv = 25k_BT$, gives $K = 10^6$ J/m^3 which represents a very significant anisotropy. In materials with lower anisotropy value, the magnetisaton will fluctuate spontaneously under the effect of thermal activation.

This effect is thought to set a limit for magnetic recording density and this limit will be reached before the year 2005. Various approaches are being examined to push superparamagnetsm to higher temperature values. These include the growth of pillars, such that the particle volume is increased but the storage density is not affected. In other approaches, the ferromagnetic film, in which the information is stored, is antiferromagetically coupled to another ferromagnetic layer or the ferromagnetic particles are embedded within an antiferromagnetic matrix.

The above discussion concerned longitudinal recording. The particle magnetisation lies in the plane of the substrate Perpendicular recording is an alternative, which will allow the recording density to be pushed forward by a factor of 4 approximately. Even more promising, is the development of "quantum bits" in which each bit of information is a unique magnetic object.Approaches towards this include film deposition on structured substrates or the direct self-organisation of magnetic nanoparticles prepared by chemistry. Spectacular results in this field have been obtained with FePt nanoparticles.

IMAGING OF MICRO- AND NANOMAGNETIC STRUCTURES

There is a great interest in understanding the micromagnetic behavior of thin ferromagnetic layers, driven by their technological exploitation in magnetic field sensors and magnetic random access memry devices (MRAM). As a high storage density is a key goal in the design of such memory devices, recent research has been concentrating on the investigation of ferromagnetic elements with submicrometer lateral dimensions. For a magnetic storage cell to deine a bit (0 or 1), it must have two stable remanent magnetization states that are uniformly magnetized, i.e. form a single magnetic domain, independent of its magnetic history.

Apart from other powerful magnetic imaging techniques, we focus on two partiular techniques to investigate the magnetic domain configurations of micro- and nanomagnetic structures: Magnetic Force Microscopy (MFM) and Scanning Near-Field Optical Microscopy (SNOM).

MFM

The MFM detects the magnetic stray fields (blue arows in Figure 1) which are generated at the edges of magnetic elements (bits) or in domain walls. For this purpose a small magnetic needle (tip) which is magnetized along the needle axis is attached to the end of a cantilever. This tip is being attractedor repelled by the magnetic stray fields pointing into or out of the surface of the magnetic element (bit), respectively.

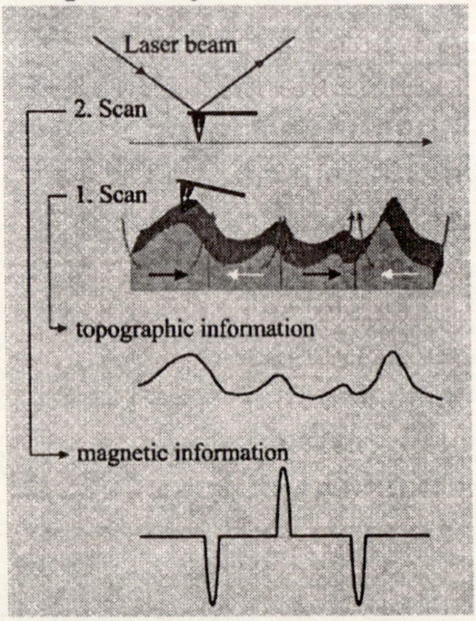

Fig 1 Imaging mechanism of a Magnetic Force Microscope.

The deflection of the cantilever resulting from this dipole-dipole-interaction is detected by a laser beam which is reflected at the pper side of the cantilever and hits a four-segment detector. Figure 1 shows schematically how the magnetic information is extracted from this signal. To avoid a cross talk by "atomic force" information the MFM first scans the topography.

The magnetic inormation is subsequently recorded by lifting the tip to a height of approximately 90 nm and scanning along the topographic profile obtained in the first scan. This method yields the typical MFM images with "dark and bright contrast" depending on directionand strength of the magnetic stray fields.

Starting with a 25 nm thick iron film with (110) orientation these elements have been fbricated by using electron beam lithography and argon ion etching. After applying a magnetic field of 1 Tesla perpendicular to the long axis of the elements - i.e. along the magnetic hard axis - the MFM image exhibits three different remanent magneticstates for zero magnetic field.

If the bright contrast represents an attractive interaction between the tip and the stray fields the upper element in the white frame shows a single domain state (or dipole state) with the magetization pointing to the right side. According to this the magnetization of the lower element in the white frame is pointing to the right. The element in the middle shows a less pronounced magnetic contrst. This is an evidence for a multi domain state (or demagnetized state). By forming such a state the stray fields are minimized by closing the magnetic flux within the element.

SNOM

When a magnetic sample is illumnated with polarized light, it rotates the polarization of the transmitted

(Faraday effect) and reflected (Kerr effect) light by an amount proportional to its magnetization. Therefore, a pattern of differently magnetized domains produces a contrast in a plarization-sensitive magneto-optical (MO) microscope. For this microscopy to work, neither a vacuum nor atomically clean surfaces are required. It can be used in an external magnetic field to probe the dependence of the domain pattern on the field.

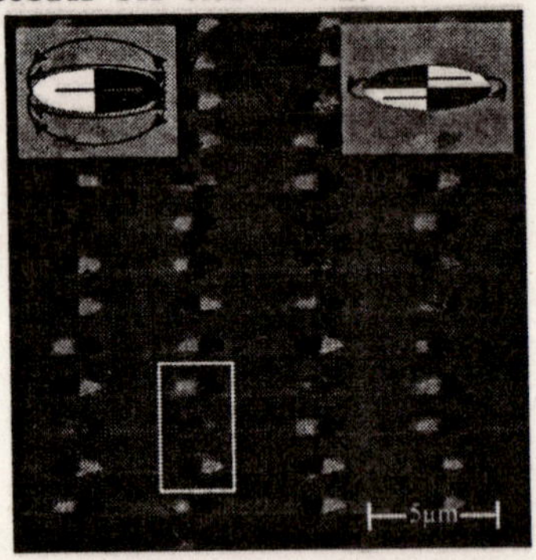

Fig 2 MFM image of microstructured ellipsoid-shaped Fe(110) elements.

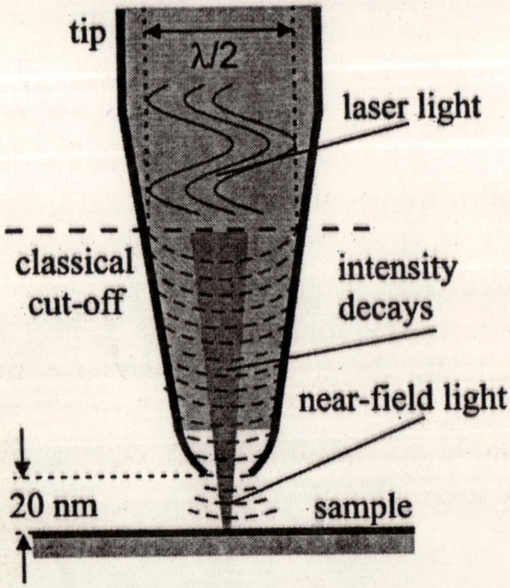

Fig 3 Imaging mechanism of a scanning near-field optical microscope.

Also, nlike in MFM, there is no danger that the domain pattern is altered by the imaging process itself. Any optical microscopy, however, is limited in its resolution by diffraction. Every illuminated spot on the sample produces its own diffraction pattern, whih cannot be distinguished from the diffraction pattern of another spot located less than $\lambda/2$ away, where λ is the light's wavelength. Smaller structures therefore cannot be resolved.

Scanning near-field optical microscopy (SNOM) eliminates the need to ditinguish diffraction patterns from one another. In the most common "aperture configuration", a probe with an aperture much smaller than l illuminates only one small spot on the sample at a time. No matter how the light is being diffracted, it i known which spot on the sample it corresponds to. A complete image of the sample is generated by scanning the probe across a grid of spots and assembling the information from all spots by computer.

How is it that light can be emitted from such a small aerture at all? Isn't there a cut-off diameter of $\lambda/2$ below which an aperture is not transparent? That depends on how close to the aperture you look. Light is actually transmitted through the aperture, but it is strongly (exponentially) damped as soon as te tip gets narrower than $\lambda/2$. Its intensity decreases by a factor of $1/e$ every 20 nm. The light is therefore called evanescent or near-field light, and for it to interact with the sample at all, the aperture must be brought to within 20 nm of the sample srface. Part of the light is converted back into normal, propagating light by the sample and can therefore be detected. If polarized light is used (MO-SNOM), differently magnetized domains produce a contrast in the image as usual.

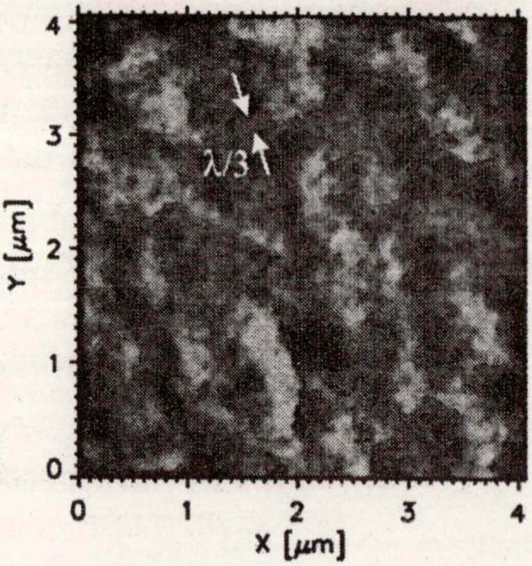

Fig 4 Magneto-optical SNOM image of a Co/Pt multilayer film.

Figure 4 shows a magnetooptical image produced with our magneto-optical SNOM set-up on a Co/Pt multilayer film. The magnetization of such a film is stable when it is perpendicular to the film plane, pointing either out of it ("up") or into it ("down").

Before the image wastaken, the film was exposed to a high (2 T) magnetic field in the film plane. When the field was switched off, there was nothing to keep the magnetization in this unstable orientation, so it relaxed into a random pattern of domains magnetized either "up" r "down". These domains appear as bright and dark areas in the image; the domain boundaries are marked by red lines. Rotating the detection set-up reversed the image contrast, proving its magneto-optical origin. The smallest domains visible in Figure 4 ae about l/3 in size, using an aperture of about 100 nm in diameter and laser light with wavelength l=488 nm.

MAGNETORESISTIVE BIOCHIPS

Detection of biomolecular recognition has been playing an ever important role when aplied to DNA-DNA hybridisation (genetic disease diagnostic, mutation detection, gene expression quantification) and antibody-antigen interaction (micro-organism detection, biological warfare agent detection, etc...). A typical DNA biochip will consist of an rray of probes (for example gene specific oligonucleotides that were immobilized onto the functionalised chip surface through microspotting), an hybridisation chamber (normally a microfluidic channel arrangement) together with an optional target arraying echanism (electric fields for charged molecules such as DNA), the target biomolecules (e.g, a complementary DNA strand to the immobilized DNA probe), the label (a fluorescent molecule that can be attached to the target), and a hybridization detection mechaism that can be either integrated on the chip or external to the chip (for instance, in the case of fluorophore labels, detection is done by an external laser-based fluorescence scanner). An hybridisation detection experiment occurs thrugh four phases, respectively 1) probe immobilization on chip surface, 2) target labelling, 3) target arraying, hybridization and washing, 4) detection. Requirements on the detection scheme depend on the particular biological assay.

For *gene expression* hips, where the relative amount of a certain gene must be quantified and compared between different patient samples, the detection scheme must lead not only to the knowledge of the presence of the particular gene (Yes or No) but also to a quantitative anaysis (present in what percentage).

For *Single Nucleotide Polymorphism* chips (SNP), where single DNA base pair changes are being identified, the simple (Yes or No) answer is sufficient. In both cases, the chip should be able to discriminate against false psitives (non-specifically bound molecules). A recent example of this technology are DNA micro-array chips fabricated by Nanogen, where charged target biomolecules are moved over immobilized probes. Electric fields are used for hybridization enhancemen and stringency control. Detection is external to the chip, done by a laser-based fluorescence scanner.

Fluorescence-based systems suffer from gradual loss of label fluorescent emission upon light excitation (photo-bleaching), and require careful backgroud signal subtraction. Alternative approaches are being pursued in various labs to incorporate other labels (or no labels at all), in order to fully integrate the detection mechanism, aiming at a fully electronic and microfluidic portable and cheap apparats, of widespread use.

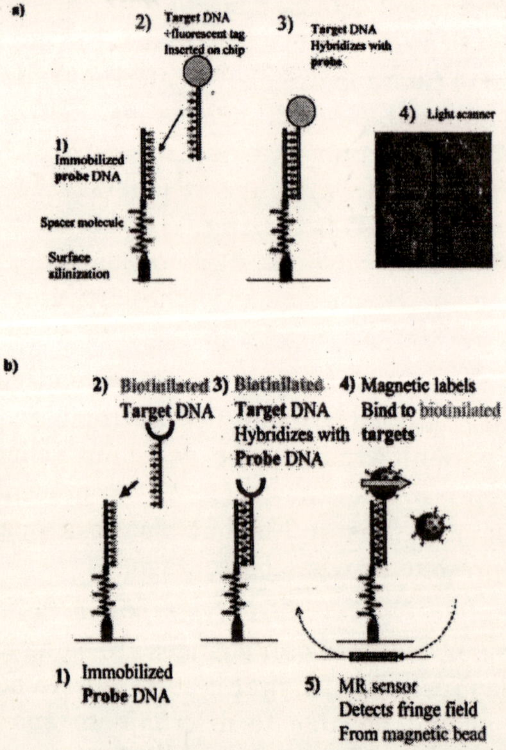

Fig 1 a) Conventional hybridisation detection using fluorescent labels attached to biological targets, and an external laser-based fluorescent scanner for detection.b) magnetoresistive-based hybridization detection using magnetic labels, and an integrated magnetoresistive sensor array for detection.

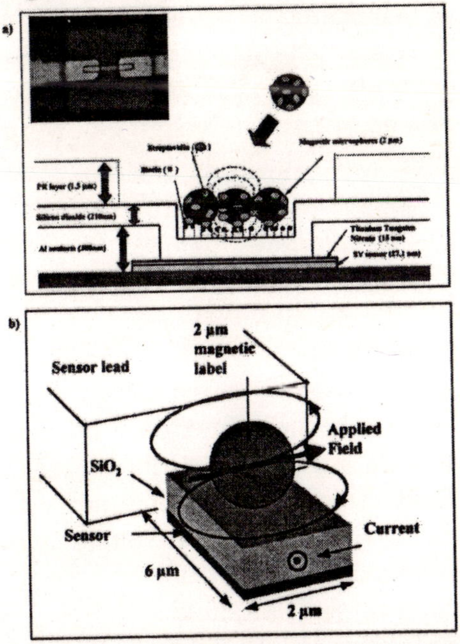

Fig 2 a) Cross-section of a MR-biochip, showing the spin valve sensor, the leads, the 2000Å thick SiO_2 functionalized passivation layer, the immobilized probes (in this case biotin), and the hybridized targets (streptavidin) coating the magnetic labels.The inset shows the top-view of the probe-pad. b) Detection geometry: a 15Oe in-plane field magnetizes the superparamagnetic labels, and in turn, these produce a transverse inplane field in the spin valve).

MR-Biochips

Magnetoresistive-based biochips were first introduced in 1998. Fluorophore labels are replaced by magnetic labels (super-paramagnetic particles), and detection is done using an integrated magnetic field sensor (GM multilayer, spin valve, AMR ring, Hall effect cross).

Since the magnetic labels used so far have been relatively large (from 100nm up to 2 μm in diameter),and may hinder the hybidization process if previously attached to the targets, hybridization occurs between the immobilized probe and a biotinylated complementary target (no label, step 3). Streptavidin coated labels are added in a post-hybridization step (4), and detection isdone in real-time. The use of magnetic labels allows the use of magnetic fields for stringency control, as well as for arraying (if previously attached to targets). Since magnetic material is usually not present in biomolecules, background signal subtracton is greatly simplified.

The MR sensor is passivated by a 0.2μm oxide or nitride layer (required to prevent sensor damage by salt solutions during immobilization, hybridization nd washing). The MR sensor directly measures the in-plane transverse field created by the magnetized label. For applied fields of 15 Oe, 2um Micromod particles (15% FeOx content) have a moment of 10^{-12} emu, creating a maximum transverse bead field of abut 1 Oe on the sensor. Fig.2b) shows the detection geometry when using an in-plane field to magnetize the beads, and spin-valve sensors for detection. Depending on the biological assay, the sensor architecture must be designed for its particular applicatin. For example consider μm size labels. If a linear response to the number of labels (up to few hundred) is required, as for example in a *gene expression* chip, the sensor can be made of a spin valve or GMR material in a meander configuration occupying mos of the area under the probe pad.

The output is proportional to the area of the sensor excited by the particles fringe fields. For a spin valve sensor, the linear range can be easily tuned from 10 Oe to few hundred Oe either by reducing sensorheight and increasing the demagnetising field, or implementing a longitudinal exchange bias field H_{bias} onto the free layer. The dynamic range for these sensors allows measurements from few to few hundred particles. If on the other hand, a SNP chip is envsaged, the spin valve sensor dimension can be tuned to that of the particle, and single labels with moments as low 10^{-14} emu should be detected.

This minimum detectable moment (field) is limited by the sensor noise, and measuring electronics setup (D vs lock-in detection). MR sensors are normally used in a Wheatstone bridge configuration in order to minimize thermal offsets and to null bridge response in the absence of particles. Fig.3a) shows a typical SNP spin valve sensor biochip, emphasizing thedifferential detection scheme, while Fig.3b) shows the picture of a packaged device, ready for biological assaying.

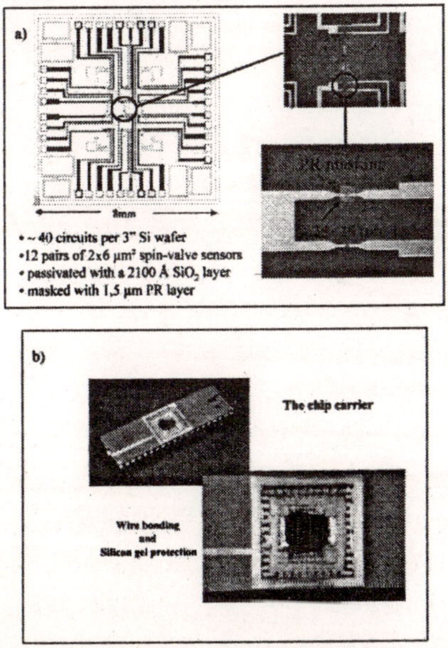

Fig 3 a) Schematics of one of the MR biochips, using differential detection, with one active sensor and one reference sensor as two arms of a Wheatstone bridge configuration. b) Packaged MR biochips ready for biological assaying.

Fig.4 Shows detection signals for few (1-4) 2μm particles over a 6X2μm^4 spin valve sensor with an MR signal of 8%. For sensor linearizatin, the sense current is set at 8mA.

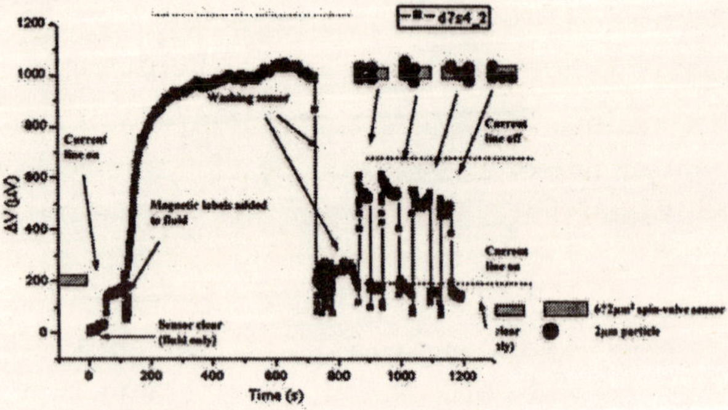

Fig 4 Bead detection using spin valve sensors.After sensor saturation (1mV), beads are washed away, and repeatidly brought over the sensor.Number of beads is checked by optical microscopy.The sensor signal reproduces clearly the bead immobilization onto and release from the sensor.

An in-plane transverse field of 15Oe is used to magnetize the superparamagnetic beads. A simple DC method was used to measure the sensor output(10μV noise level). For this experiment, the labels were controllably move across the sensor using field gradients created by auxiliary current lines. Typical signals for single beads are of the order of 150uV/bead. DNA-cDNA hybridization detection: cystic fibrosis CFTR gene recognition

As an example of DNA-cDNA hybridizationrecognition using the MR biochip a cystic fibrosis specific 50mer (50 base pairs) oligonucleotide corresponding to a particular strand of the CFTR gene was immobilized onto the 2000 Å thick, sputtered SiO_2 probe pad, covering the MR sensor area. The goal s to detect whether a complementary target will bind or not. Prior to probe immobilization, SiO_2 functionalization was achieved by surface silinization followed by a cross linker placement. Hybridization was attempted using either complementary or non-comlementary biotinylated DNA target sequences.

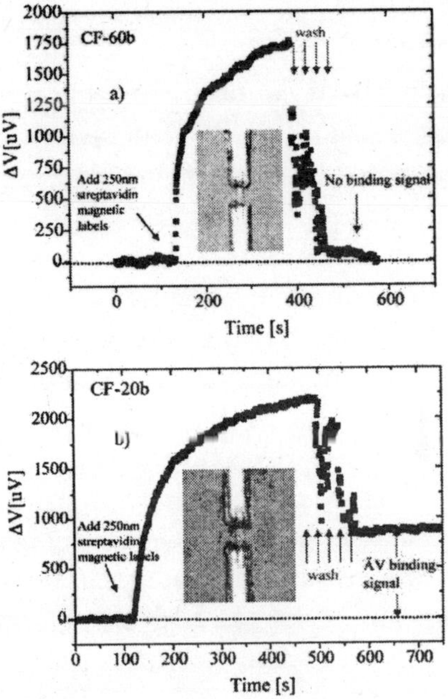

Fig 5 a) Detection of DNA recognition using streptavidin coated 250nm labels, and a biotinylated complementary target.After washing to eliminate nonspecifically bound molecules, a strong remnant signal remains, from the bound complementary targets. b) When a non-complementary target is used, after washing, no remnant signal is observed.

After hybridization and washing to remove non-specifically bound DNA targets, streptavidin functionalized magnetic labels are introduced in solution. These attach to the bound, biotinilated complementary DNA tagets. Real time measurement of the sensor output during multiple label injection and washing cycles show that where hybridization has occurred, MR sensor remnant signals of 1mV are observed, corresponding to about 100 nanoparticles (250nm) bond. Direct optical analysis confirms the MR signal results. When a non-complementary DNA strand is used, no remnant signal is observed.

CONCLUSION

MR technology is being successfully applied to biomolecular recognition in different contexts. Advnces are required in the development of biocompatible magnetic labels with higher moments (eg, pure Co, Fe, NiFe particles), and nanometric dimensions (< 100nm) providing clustering can be avoided and biocompatibility assured. Developments in the sensing echnology are required to allow measurements of single nm-sized labels with moments at the 10^{-15} emu level. Developments are required in the use of magnetic and electric fields for arraying, hybridization enhancement, and force discrimination of non-speciically bound biomolecules. MR technology has shown the potential for single molecule process detection, a target not usually within the reach of most of the competing technologies.

9

The New Era of Spintronics

INTRODUCTION

Spintronics, at the interface between magnetism and electronics, is a new field of research in considerable expansion. The basic concept of spintronics is the manipulation of spin currents, in contrast to mainstream electronics in which the spin of the electron is ignored. Adding the spin degree of freedom provides new effects, new capabilities and new functionalities. Beyond today's applications to hard discs and memories, the potential of spintronics is very promising for new advances and an important impact on science and technology of the 21st century.

The influence of the spin on the electronic transport properties of ferromagnetic materials has been known for a long time. The existence of spin polarized currents in ferromagnetic metals, first suggested by Mott, has been experimentally demonstrated and theoretically described about thirty years ago, mainly by European researchers (at Orsay, Strasbourg, Eindhoven).

In metals like iron or cobalt, the "majority spin" and "minority spin" electrons carry currents that can be different by more than a factor of ten. This spin dependent conduction comes from the splitting of the "majority spin" and "minority spin" energy bands and also from the spin dependent cross section of some defects or impurities.

Everybody has Already a Spintronic Device on their Desktop

The first step on the road of the utilization of the spin degree of freedom was the discovery of the Giant Magnetoresistance of magnetic multilayers in 1988. A magnetic multilayer is a film composed of alternate ferromagnetic and nonmagnetic layers, Fe and Cr for example. The resistance of such a multilayer is lowest when the magnetic moments of ferromagnetic layers are aligned, and highest when they are antiparallel. As the relative change of resistance can be as high as 200%, this effect has been called Giant Magnetoresistance (GMR). The condition required for GMR is a much better conduction through a ferromagnetic layer by the electrons of one of the spin directions, say the majority spin direction for example.

When the magnetic moments of all the layers are aligned, half of the electrons are majority spin electrons in all the magnetic layers, and the short circuit effect by this high conduction channel results in a low resistance. In the antiparallel configuration, each electron is alternately a majority and a minority spin electron, the short circuit effect does not exist and the resistance is much higher. In specially designed multilayers, known as spin valves, the magnetic configuration can be switched between parallel and antiparallel configuration by a field of only a few Oersted, so that a large change of resistance can be induced by a very small field.

The first spintronic devices have been based on spin valves. Today everybody has already a spintronic device on their desktop, since all modern computers use spin valves for the read heads of the hard disc. Because they can detect very small fields and very small magnetic bits, the spin valve-based read heads have led to an increase of the density of stored information by almost two orders of magnitude. It might however be hard to go beyond the 100 Gbit/in^2 of today's prototypes with conventional spin valves. A further increase of read head sensitivity and information density in hard discs will be probably achieved with other spintronic devices, magnetic tunnel junctions (see next paragraph), patterned multilayers for GMR in the non-conventional geometry where the current is perpendicular to the layers or magnetic nanocontacts presenting a ballistic version of GMR known as BMR.

Magnetic Tunnel Junctions and MRAM

The magnetic tunnel junction (MTJ) is the second type of spintronic device that will have soon important applications. A MTJ (Fig. 1a) is a structure in which two ferromagnetic layers (electrodes) are separated by a very thin insulating layer, commonly aluminum oxide. The electrons can tunnel through the insulating layer and, because the probability of tunneling from a ferromagnetic electrode depends on the spin direction, the resistance of the MTJ is different for the parallel and antiparallel orientations of the magnetic moments of the electrodes.

For electrodes of conventional ferromagnetic alloys, the relative change of resistance (Tunnel Magnetoresistance or TMR) can reach 70% at room temperature. MTJ of very small size, below the micron range, can be fabricated by lithographic techniques and an important application of these small size MTJ will be for a new type of computer memory, the MRAM (Magnetic Random Access Memory). As illustrated in the upper part of Fig.1, each junction can store one bit of data, say "0" for the parallel configuration of the magnetic moments of the electrodes, "1" for the antiparallel configuration.

The sketch below is a schematic representation of a MRAM with MTJ connected together in a point contact array. The MRAM presently in development are expected to reach similar densities and access times as the current DRAM or SRAM, but their main advantage on these volatile semiconductor-based memories is that they retain data after the power is turned off, possibly eliminating the long boot-up time when the computer is switched on. The MRAM have also an important advantage in terms of speed (a gain of 1000 for the write time) over the permanent memories of Flash type that are used today for mobile electronics. At least two commercial efforts-one by Motorola, the other by IBM and Infineon in Munich-plan to begin production of MRAM in 2004.

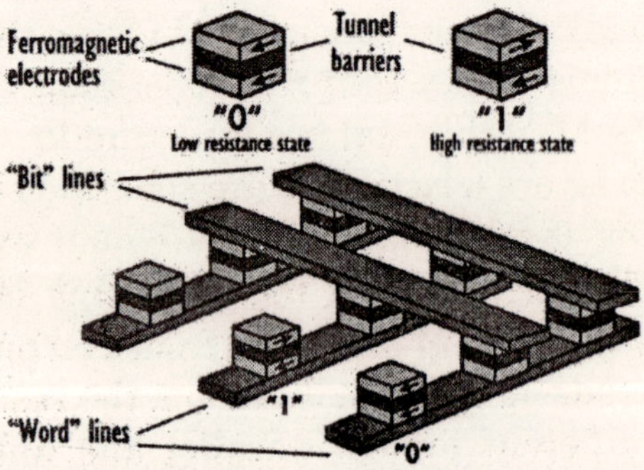

Fig 1 Top:Memory cells of an MRAM (Magnetic Random Access Memory). Each cell is a submicronic MTJ and the states "0"and "1"of the cell corresponds respectively to the parallel and antiparallel configurations of the magnetic moments of electrodes of the MTJ (low and high resistance states).Central sketch:schematic of an MRAM constructed of MTJ connected together in a point contact array.Conducting wires,"bit lines"and "word lines"permit voltage measurements to read the stored information.They also enable manipulation of the magnetization of the elements (writing) by the magnetic field created by currents in the lines.The main advantage of MRAM over semiconductor-based DRAM or SRAM is their permanent character (for similar density and speed).

Though commercial products using MTJ are expected very soon, the physics of spin dependent tunneling is still far from being clearly understood. In the early research on the MTJ, it was thought that the spin polarization of the tunneling electrons and the TMR were simply reflecting the spin polarization of the electronic density of states at the Fermi level in the ferromagnetic electrodes. It now turns out that the TMR also depends on the insulating material of the tunnel barrier and, more specifically, on the details of the electronic structure at the electrode/barrier interface.

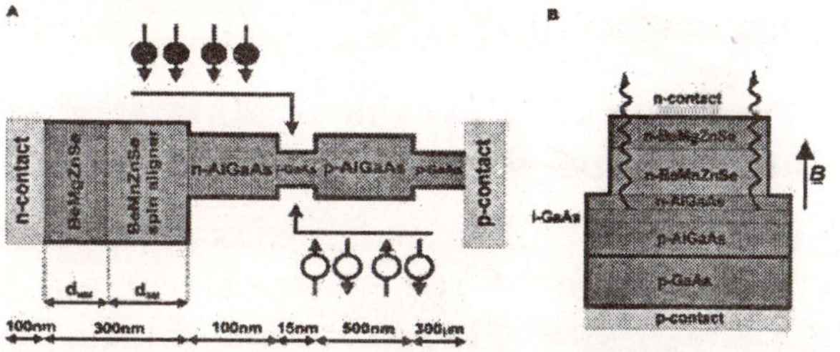

Fig 2 Spin Light Emitting Diode (Spin- LED).A current of spin-polarized electrons is injected from the diluted magnetic semiconductor $Be_xMn_yZn_{1-x-y}Se$ into a GaAs/GaAlAs light-emitting diode A circularly polarized light is emitted from the recombination of the spin polarized electrons with non-polarized holes. An injection efficiency of 90% spin polarized current has been demonstrated by Fiederling *et al*..As $Be_xMn_yZn_{1-x-y}Se$ is paramagnetic, the spin polarization is obtained only in an applied field and at low temperature.

The present theories are not really quantitative yet and further theoretical developments are still needed for more reliable predictions. Another important challenge is the research of ferromagnetic materials providing higher spin polarizations than conventional metals like cobalt or iron. A few ferromagnets have been predicted to be half-metallic, that is to present a spin polarization of 100% at their Fermi level. A record TMR ratio of 1800%, corresponding to a spin polarization of 95%, has been found with electrodes of the manganite $La_{2/3}Sr_{1/3}MnO_3$ (LSMO).

However these results are obtained only at low temperature and LSMO cannot be considered for practical devices. Better candidates for room temperature applications seem to be magnetite (Fe_3O_4) and several oxides of double perovskite structure. High TMR ratios might also be obtained with ferromagnetic insulating materials presenting a spin dependent gap and acting as spin filter for the tunneling electrons. Some interesting results have been recently obtained with EuS barriers, but at low temperature only.

Spin Transfer to Reverse a Magnetic Moment without Applying a Magnetic Field

In a GMR or TMR device, switching the magnetic configuration changes the electrical current passing through the device. Magnetization reversal by spin transfer is an opposite effect: a spin polarized current is let into a device and the transfer of spin from the current switches the magnetic configuration of the device.

This concept of magnetization reversal by spin transfer from a spin-polarized current has been introduced by Slonczewski in 1995 and has been now confirmed by series of experiments (mainly on pillar-shaped multilayers). From the application point of view, reversing a magnetic moment by spin transfer without applying an external applied field can be of great interest to switch spintronic devices, MRAM for example. However the current density needed in the existing experiments is still relatively high, of the order of $10^7 A/cm^2$, and a better understanding of the spin transfer mechanisms seems necessary to obtain a significant reduction ot the current density. Another effect of the same type but probably requiring a smaller current density is the displacement of a domain wall by spin transfer from a spin-polarized current.

SPINTRONICS WITH SEMICONDUCTORS

Whereas the metallic spin devices just described provide new ways to store and read information in hard discs, tapes or MRAM, semiconductor-based spintronics may offer a greater wealth of possibilities. Why is spintronics with semiconductors interesting? First, semiconductor-based spintronics could combine storage, detection, logic and communication capabilities on a single chip to produce a multifunctional device that could replace several components.

For example, it could permit a better integration between MTJ and silicon-based electronics than in the present prototypes of MRAM. The optical properties of the semiconductors are also of particular interest to transform a magnetic information into an optical signal. Finally, because the manipulation of spins presents some advantages in term of speed and required power over the manipulation of charge in conventional electronics, some concepts of device exploiting these advantages have been already proposed.

A first interesting property of the semiconductors for spintronics is the long electron spin lifetime. Fine time-resolved magneto-optical experiments have been developed at the University of Santa Barbara and have revealed spin lifetimes that can exceed 100 ns for electrons in bulk semiconductors at low temperature. In heterostructures and quantum dots, nanosecond dynamics persist at room temperature. In most of these experiments on spin dynamics, the spin polarization is created by optical excitation. However, for practical applications, it is highly desirable that the injection and detection of spin currents be electrical.

The most direct way for spin injection would seem injecting from a classical ferromagnetic metal in a metal/ semiconductor heterostructure but this raises difficult problems related to the difference in conductivity and spin relaxation time in metals and semiconductors. Although these problems are now better understood, this has slowed down the progress for spin injection from metals. On the other hand, this has boosted the research on diluted magnetic semiconductors (DMS) that can be associated more easily with nonmagnetic semiconductors for spin injection. GaAs doped with Mn is an archetypal DMS which presents ferromagnetic properties up to 150 K. Some other compounds are promising for room temperature ferromagnetism.

We will describe two examples of spin injection from DMS in experiments respectively achieved at the University of Wurzburg and in our team.

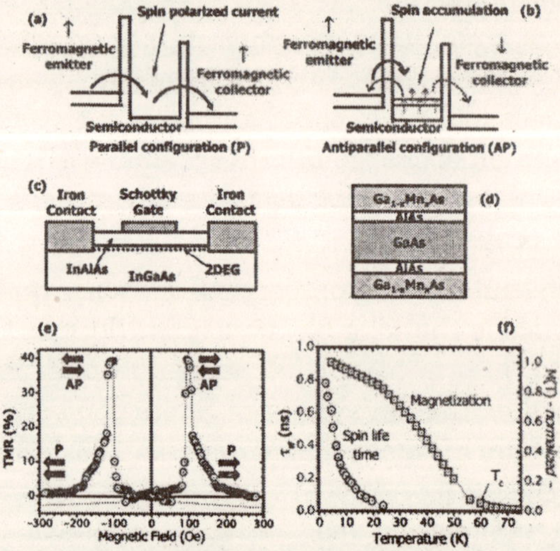

Fig 3 (a,b) Sketches illustrating spin injection from a magnetic spin emitter (metal or semiconductor) into a semiconductor (2Delectron gas channel,Quantum Well, etc.) and spin detection by a magnetic collector (spin analyser). Injection and detection are through tunnel barriers. For a parallel (P) configuration of magnetic moments in the emitter and collector (a), a spin-polarized current (spin up on the figure) is injected and transmitted to the collector. For an antiparallel (AP) configuration (b), spin up electrons are injected and accumulated in the semiconductor (accumulation due to the poor transmission to the collector) and rejected in major part by the spin up accumulation into the emitter.The condition for strong accumulation and large difference between the resistances of the P and AP configurations is that the spin lifetime in the semiconductor be longer than the time spent by the particle in the semiconductor. (c) Lateral heterostructure with metallic spin emitter and spin analyser for spin injection into a 2D Electron Gas (DEG).With a Schottky gate to rotate the spin polarization by the Rashba effect, this device is the Spin Field Effect Transistor proposed by Datta and Das. (d) Vertical heterostructure for injection of spin-polarized holes into a GaAs Quantum Well (QW) in the experiments of Mattana *et al.*. Emitter and analyser are layers of the ferromagnetic semiconductor $Ga_{1-x}Mn_xAs$. (e) Magnetoresistance of a heterostructure of the type of (d). (f) Temperature dependence of the spin life time in a GaAs QW from magnetoresistnace measurements on a heterostructures of the type of (d).

In the heterostructure of Fig.2, spin-polarized electrons are injected from a paramagnetic DMS ($Be_xMn_yZn_{1-x-y}Se$) into a GaAs/AlGaAs light emitting diode (LED), which leads to emission of circularly polarized light. An injection efficiency of 90% spin polarized current has been demonstrated with this structure. Similar experiments but with smaller polarization have also been performed with spin injection from ferromagnetic GaMnAs and from metals.

Fig.3 illustrates the physics of devices where both injection of spins into the semiconductor and detection of the spin information are electrical. The ideal situation is when the spin lifetime is much longer than the time spent by the carriers in the semiconductor. A spin-polarized current is then easily transmitted in the parallel configuration of emitter and collector, whereas the antiparallel one leads to spin accumulation and current blockade. In the Datta and Das spin transistor of Fig.3(c), a gate voltage can rotate the spins to switch the device between the two regimes. Fig.3(d) represents an example of vertical version of such an emitter/semiconductor/analyser structure, and, in Fig.3(e), we show the magnetoresistance resulting from the coherent transmission of the spin information in this device. The temperature dependence of the spin lifetime in the GaAs QW derived from this magnetoresistance is shown in Fig.3(f).

The two examples of Fig.2 and 3 show it is now possible to inject spins electrically into a semiconductor heterostructure and to transform the spin information into an optical (Fig.2) or electrical signal (Fig.3). This paves the way for more advanced spintronic devices in which it will be possible to manipulate the spin polarization between emitter and analyser.

In just a dozen of years, we have seen spintronics increasing considerably the capacity of hard disks and now getting ready to enter the RAM of computers. In the next decade, spintronics with semiconductors has the potential to gain an important place in the microelectronics industry. Another perspective, at longer term and out of the scope of this paper, is the exploitation of the truly quantum-mechanical nature of spin and the long spin-coherence time in confined geometry for quantum computing in an even more revolutionary application.

MAGNETISM IN NANOCRYSTALS

Physics of magnetic materials prtends the improvement of the magnetic properties useful for applications. These can be roughly classified in two big groups, i) magnetic flux multiplication (magnetic cores) and ii) magnetic storage of either energy (magnets) or information (magneticrecording). There also exist other sophisticated devices as sensors and actuators that, being based on magnetic properties of materials as magnetostriction or giant magnetoresistance, GMR, are widely used in electronic and microelectronics circuits.

Fr flux multiplication soft magnetic materials characterised by low losses and coercivity, Hc, are required; whereas, high Hc and energy product are the relevant properties of hard magnetic materials used for storage. Control of coercivity has been carriedout by playing with the parameters that governs the magnetic anisotropy i.e. composition and structure. The possibility of tailoring anisotropy has allowed us to spread nowadays the available spectrum of coercivities over seven orders of magnitude.

Bu what is really remarkable is that the softest materials Fe79Zr7B9 (μ0Hc=10-7T) as well as the hardest material Fe79Nd7B9 (μ0Hc=10T) known in 2000 are obtained from amorphous alloys with closely related compositions. Notice that only a difference of 7 at.% in content gives rise to a difference of seven orders of magnitude in coercivity. This enormous difference is due to the nanocrystalline structure obtained by partial devitrification of the initial amorphous state. Both types of samples consist of a sou of nanograins embedded in a softer matrix. Therefore, it can be concluded that nanocrystalline materials form an outstanding group of magnetic materials as concerns applied magnetism.

In this review the basic definitions and classification of nnocrystalline materials as well as their more remarkable intrinsic characteristic are introduced. Finally two examples of nanocrystalline magnetism related to pure Fe ferromagnetic grains are described. The first example deals with the amazing magnetic beaviour observed at low temperature in nanocrystalline Fe and the second one is related to the anomalous thermal dependence of the spontaneous magnetization in nanosized Fe ribbons embedded in a Cu matrix. It is shown that such anomalous thermal dependencecan be explained by taking into account the combined effect of size and shape on the spin wave thermal excitation. These two examples are expected to be useful to illustrate the relevance of nanocrystals not only on technical magnetism but also on more fudamental aspects of magnetism.

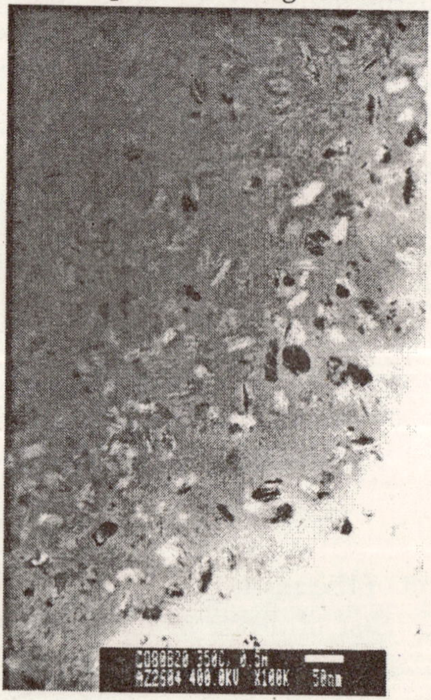

Fig 1 Nanocrystalline system formed by Co nanograins embedded in a Co rich CoB amorphous ribbon as observed by TEM (Zern *et al.*).

A polycrystalline solid formed by an isotropic distribution of nanometric sized grains is known as a nanocrystalline material. Figure 1 shows an assembly of

Co nanograins embedded in an Co rich amorphous matrix. The structue corresponds to the partial devitrification of amorphous $Co_{80}B_{20}$ obtained after annealing at 630 K for 30 minutes. In general nanocrystals are connected between them by either grain boundaries or a different matrix. In the case of ferromagnetic nanocrystls, the matrix can be magnetic or non magnetic. The strength of the exchange coupling between the magnetization of the adjacent ferromagnetic grains strongly depends on the average thickness, d, and the magnetic nature of the matrix. It is important to reark that the macroscopic properties of nanocrystalline ferromagnetic samples are governed by the nanometric dimensions of the grains and also by the degree of ferromagnetic coupling between grains. Hence, the matrix nature becomes an extremely important prameter.

The relevance of nanocrystals in magnetic research is associated with the control of the magnetic and magneto-transport (magnetorresistance and tunnel magnetorresistance) properties. Before going into deeper details, let us state the tw outstanding physical aspects of nanocrystalline materials. First, is the high value of the ratio, N_s/N_v, between the atoms at the interface, N_s, and the total number of atoms N_v. If we call D the average grain size and a the inter-atomic distance, N_s/N_v aries roughly as a/D that, when D is a nanometric distance, can reach values as high as 30%. Second, there is coincidence between the grain size scale with the exchange correlation length, L_0, or domain wall width of the bulk ferromagnetic materials with he same composition as that of the nanocrystals. In fact L_0 is the distance at which any local disturbance in the collinear ferromagnetic order propagates through the material.

The actual exchange correlation length in the nanocrystalline sample, L,can be either larger or shorter than L_0, depending on the magnetic nature of the matrix interconnecting the grain. Let us restrict the discussion to those nanocrystalline materials with grain size D very close to L_0 and average distance between grains, d f the order or smaller than D. The following two cases, described in Figure 2, can be distinguished

i) *First Case:* L<D.This happens always for L_0<D,as, for example, is the case of nanocrystalline $Fe_{79}Nd_7B_9$.But, even if L_0>D, it is also possible that L<D if the matrix does not transmit exchange between grains (non magnetic matrix). For this case the grains are ferromagnetically decoupled and should closely behave as isolated magnetic single domains. In general, the systems for which L<D would exhibit high temperature superparamagnetism, low temperature hard magnetic behaviour and giant magnetorresistance,GMR.The particles are only coupled through magnetostatic interactions whose strength depends on the particle dilution or average intergrain distance, d.

ii) L > D. This case only holds under two independent conditions, a) L_0>D and b) the matrix is a good exchange transmitter (soft ferromagnetic matrix). In this case all the grains contained in L^3 are exchange coupled among them. The magnetization process is collective and the average anisotropy falls from the anisotropy of an isolated nanograin k down to $k^* = k/N^{1/2}$.Under the alignment effect of the exchange interaction between grains the magnetic moments tend to arrange in a ferromagnetic structure. Since the easy axis fluctuates in orientation from one grain to the adjacent (random anisotropy), the exchange effect is counterbalanced by the misalignment tendency exerted by the anisotropy at each nanograin.When the exchange is stronger than the local anisotropy, k, the average anisotropy, k*, decreases as expected from random walk considerations. Therefore, it can be concluded that the extremely soft magnetic character of nanocrystalline Fe79Zr7B9 is due tothe effect of random anistropy.

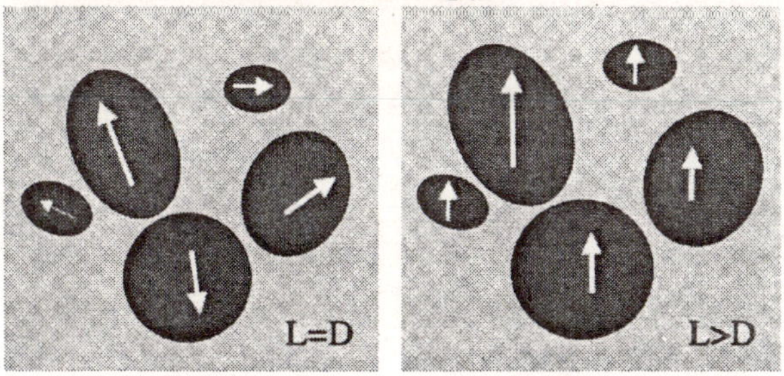

Fig 2 a) AFM image of the Fe grains; b) Electrical resistivity measured during zero-field cooling and field cooling.The dramatic increase of resistivity observed at low T in the zero-field cooling curve is originated by the disorder spin structure frozen in the grain boundaries.

In order to illustrate the influence of the nanostructure in the magnetic properties pure iron has been found to be an excellent example. In fact it is somehow amazing that Fe, the oldest known ferromagnetic metal that has been considered the arhetype of a ferromagnetic material, exhibits new and singular magnetic properties when it is in the form of nanostructures. Indeed, recent experiments performed on nanocrystalline Fe systems obtained by different methods have shown that intrinsic propertis, such as the thermal dependence of the spontaneous magnetization, and technical aspects, as magnetic softness, are strongly modified by the nanometric scale of the Fe grains.

Let us start with the case of pure nanocrystalline Fe in which the contact btween grains takes place through the grain boundaries that act as matrix. After measuring the thermal dependence of the magnetization under weak applied fields after field cooling, FC, and zero field cooling, ZFC, processes, a dramatic hardening of the saples was observed for temperatures below 70 K. The hardening was detected by the decrease of the real component of the susceptibility and the dramatic drop of the thermoremanence. It is important to remark that this low temperature anomaly has been observd in nanocrystalline Fe obtained by two very different methods such as ball milling and sputtering. Mössbauer spectroscopy carried out on strongly milled Fe powders with average grain size of 12 nm indicated than 13 at% of the Fe atoms were subjcted to an hyperfine field different to that of regular a-Fe, ie. 33 T.

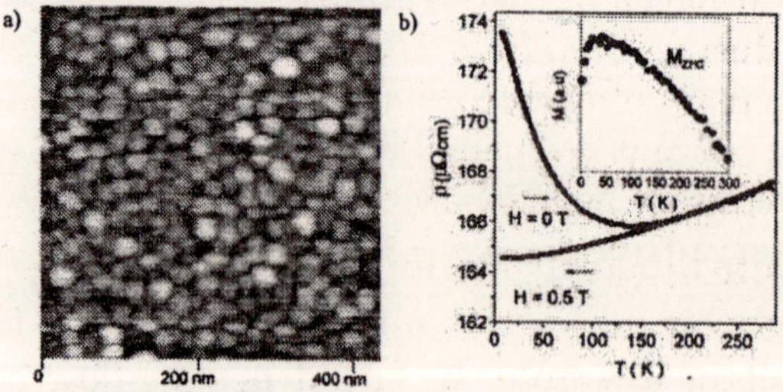

Fig 3 a) AFM image of the Fe grains; b) Electrical resistivity measured during zero-field cooling and field cooling.The dramatic increase of resistivity observed at low T in the zero-field cooling curve is originated by the disorder spin structure frozen in the grain boundaries.

This result suggested that 13 at% was the percentage of atoms located at the grain boundaries that, due to fluctuations of interatomic distances and coordination angles, exhibited a boad spectrum of hyperfine fields. Since magnetism is known to be extremely sensitive to short range order, the hardening was suggested to be due to spin disorder being frozen-in within the boundaries at low temperature. Spin disorder must drastically decrase the ferromagnetic exchange connection between grains giving rise to a shortening of L that promotes magnetic hardening. In order to confirm this suggestion electrical resistivity measurements were performed in thin films formed by (110) Fe islands gron by sputtering onto Al^2O^3 (0001) substrates in an ultrahigh vacuum triode sputtering system at 980 K.

Figure 3-a shows the structure of the film observed by Atomic Force Microscopy and Figure 3-b illustrates the thermal dependence of resistivity measuredduring field cooling and zero-field cooling processes. The enormous increase of resistivity observed at low temperature under the zero-field cooling procedure and not observed when the cooling is achieved under a field can be explained by the proposed spi disorder at the boundary that also contributes to increase the scattering with conducting electrons and therefore to enhance resistivity.

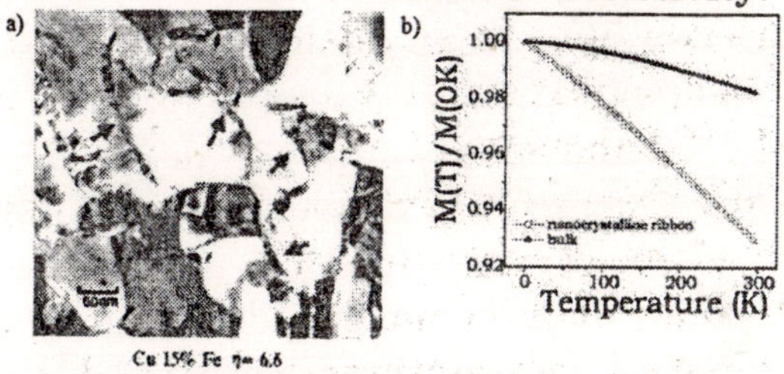

Fig 4 a) Scanning Electron Microscopy image of the composites.The nano-ribbon of Fe can be observed at the grain boundaries of Cu grain; b) Thermal dependence of the reduced magnetization for bulk and nanocrystalline ribbon shaped Fe.

Therefore, at low temperatures the exchange correlation length and the electron mean free path are remarkably reduce in nanocrystalline Fe. These effects are consequences of the low temperature disordered distribution of those spins located at the structurally disordered grain boundaries. It is obvious that this effect is not appreciable in polycrystalline Fe for whichthe percentage of atoms at the boundaries is negligible and the exchange connection between grains does not have any significant influence on the macroscopic magnetic properties.

Finally let us discuss the effect observed in Fe ribbon shaped nanocrystalsembedded in a Cu matrix. Wire Cu-15 vol%Fe composite alloys have been produced by heavy working of spray-deposited billets. With the cold working proceeding the spheroidal Fe particles, about 1 μm in size, become drawn into elongated, flattened ribbons wih progressively smaller cross section and (1,1,0) texture. Figure 4-a shows the microstructure of the composite and Figure 4-b summarises the experimental thermal dependence of the reduced spontaneous magnetization, M(T)/M(0K), for bulk Fe and that masured in Cu-Fe composites formed by Fe grains with 3.5 nm thickness, 27 nm width and 5 X 10^3 nm length.

The nanocrystalline sample exhibits an almost linear decrease of magnetization with temperature, T, instead of the usual $T^{3/2}$ bahavior predicted by th Bloch law and experimentally observed in bulk samples. This effect can be understood by considering that spin waves with wavelength larger than the nanocrystals dimensions can not be excited. At low temperatures the Planck distribution indicates that onl large wavelengths are excited to a considerable extent. Thus, in nanometric spheroidal particles the magnetization is stable at low T up to that T at which spin waves of the sample dimensions are allowed to be excited. However, in the case of ribbon shapd nanocrystals, the spin waves excited at low T can only be those with wave vector oriented along the particle length. But, as is well known, the number of spin waves excited in one dimension diverges in the limit of very large wavelength. Therefore, the umber of spin waves excited within the T interval in which all of them are one dimensional is enormous. Such T interval increases with the aspect ratio.

We have shown two examples that illustrate the sensitivity of Fe magnetism to short range order. Thespin disorder at grain boundaries of Fe nanocrystals indicates the dependence of the magnetic structure on the topological short-range order. The shape influence, at the nanometric scale, on the spin-wave excitation rate also emphasises the relevance of nnocrystalline structure for controlling the intrinsic magnetic properties of the materials.

MAGNTORESISTANCE AND MAGNETOSTRICTION IN MAGNETIC CONTACTS

Ballistc Magnetoresistance in Atomic and Naometer Size Electrodeposited Contacts

Ballistc magnetoresistance in atomic and naometer size electrodeposited contacts Ballistic magnetoresistance (BMR) in point atomic contacts with resistances (R) larger than around 1000 Ω exhibits very large values, up to 500%, while the general trend is that BMR increases when R decreases.This BMR is the resistance that the electric current manifests through a point contact when a magnetic field is applied. The cause of this phenomena is explained in terms of domain wall (DW) scattering in magnetic materials, assuming that the DW widh formed at the contact is of the order of the contact size. If the DW is narrow, strong non-adiabatic scattering with spin conservation at both sides of the contact is predicted by an old theory of Cabrera and Falicov, applied to atomic contacts by Ttara *et al.*. The BMR can be expressed by:

$$BMR(\%) = \pi^2/4(\zeta^2 /1\text{-}\zeta^2)F(\xi) \quad (1)$$

The factor F is a dynamic term that reduces BMR according to the DW width and shape, and describes the adiabaticity in the spin electron transfer from one side of the contct to the other. The rest of eq. 1 is the ratio of the difference in the density of states between majority and minority spins at the Fermi level in the electrodes at both ends of the contact. This is basically the Julliere formula for two electrodes ade of same metal. The resistance is spin ballistic if the contact is small and the spin mean free path (*l*) is larger than the contact size.

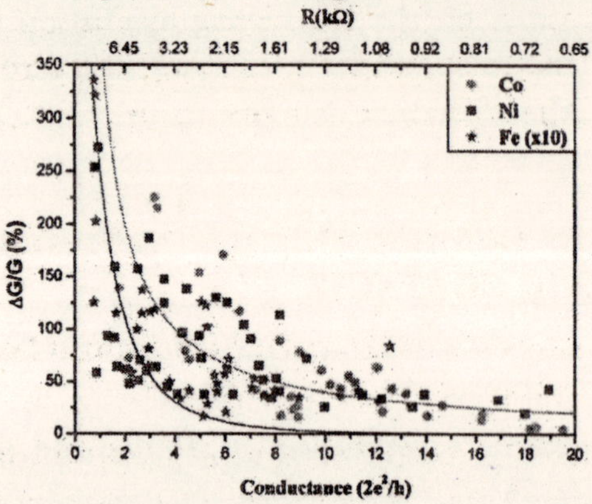

Fig 1 Magnetoconductance as a function of the contact conductance (bottom x-axis) and the contact resistance (top x-axis) for Ni, Co and Fe contacts. Black and red are the theory line approximations in the limits of a small and large number of conducting channels, respectively.

The resistance can be estimated assuming that one atom at the contact provides a conducting channel, and that oneatom takes a surface area of approximately 0.1nm^2 by:

$$R(\Omega) = 12900\ \Omega\ /\ (10\ a^2\ (\mathrm{nm}^2)) \qquad (2)$$

where 12900 Ω is the quantum of resistance, and a is the size of the contact in nm. The above formula is basically Sharvin's formula for ballistic transport The typical values from (l) for the spin mean free path are 30-50nm, and it therefore can be observed from the above formula that the limit for ballistic transport are in R values larger than 1Ω. For much smaller values of R the classical Maxwell formulacontrols the resistance that becomes non-ballistic. Fig.1 shows BMR values for Ni, Fe, and Co as a function of the conductance expressed in quantum units. The resistance of the contact is obtained simply by dividing the quantum of resistance 12900 Ω by th number of conducting channels (given in the upper x-axis). Fig.1 indicates that, for example, for 20 channels, corresponding approximately to 2 nm^2 of contact area, the resistance is 600 Ω! By then, the BMR is largely reduced to a 10% value of that for oe channel of conductance. This conforms very well with theory, as indicated by the good agreement between the lines (theory) and dots (experiment) in the Fig 1. While this observation may produce very interesting physics from the point of view of theory ad experiments, it does not qualify for technological applications because the contacts are unstable and last only a few minutes.

To overcome this problem and improve the stability of the contacts, we have electrodeposited contacts in magnetic and nonmagntic wires and films, forming a "T" configuration. To our surprise, we have observed even larger values of MR; up to 700% in Ni contacts deposited on small gaps between Ni wires with typical R-values of 5-20 Ω. These MR values clearly cannot be explaind by DW scattering, since the contact sizes are of the order of 10 nm, while the DW scattering yields BMR values not larger than 10%.

Recently, much larger values (many thousands per hundred!) of MR have also been claimed. This effect f large BMR values in 10 nm size contacts may be assigned to the formation of a very thin "dead layer" or non-stoicheometric compound (oxide, sulfite, etc), at the nanocontact region that may reconfigure the spin density of states defining the electron tansport. While the very large values are very difficult to reproduce and to stabilize, we have been able recently to obtain very stable contacts with highly reproducible R(H) curves; i.e., variations of resistance versus applied magnetic fiel H.

These last recent magnetic contacts were formed with two films of Cu as a substrate, separated by a gap between 100 and 2 μm wide that we can create with electrodeposition techniques, Fig.2a. Once the desired gap is created, we deposit the magnetic cntact. This material may be Ni, Fe, Co or whatever as well as binary compounds such as Permalloy $Ni_{85}Fe_{15}$, which is very soft and has a small coercive field. With this procedure we have obtained extremely stable and reproducible R(H) curves that can last or the duration of thousand of loops. As an example, Fig.2b shows the reproducibility of the R(H) curves, for only some tens of the hundreds of curves that have been measured. The reproducibility is astonishing and is encouraging for the development of hihly sensitive sensors as magnetic reading heads. At present we can obtain these kinds of results with a yield of 60% out of the thousands of samples we have studied. The BMR values, obtained with the typical reproducibility given in fig.2(b), range betwee 20% and 300% for contacts of the order of 10 Ω that correspond to contacts of approximately10nm in size.

Magnetoresistance of the Deposit Controls its Morphology and Structure

To prove the magnetoresistive effect we have performed electrochemical experients applying a magnetic field during deposition. The idea goes as follows:

a) We grow Ni by electrodeposition with zero applied fields at several anodic potentials.

b) We, subsequently, make the same experiments with an applied magnetic field using wea fields up to 500 Oe. This field is approximately the coercive field of the Ni electrodeposits as we observed in our BMR experiments on nanocontacts. This is not, however, to be confused with Fig.2(b) where the coercive fields are much smallerbecause in the latter case the deposit is permalloy and has a 25 Oe coercive field.

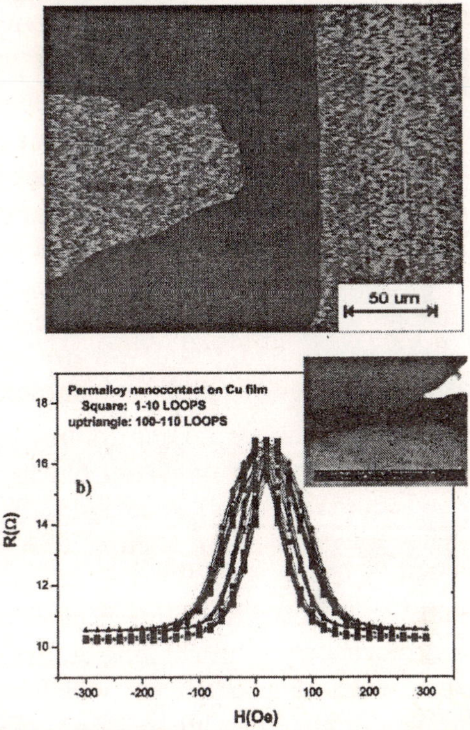

Fig 2 a) Optical microscope photograph taken near the contact area of the "T" configured Cu film arrangement. b) BMR curves of deposited permalloy contacts on Cu film.We measured the sample twice, taking one hundred loops per measurement. Indicated here are the first ten loops of each measurement: the upward-pointing triangular data points depict loops 1 through 10, while the square data points depict loops 100 through 110. The inserts at the top right of the curve depict SEM scans corresponding to the contacts referred to with the scales shown at the bottom of the photos.

c) The field is applied in the direction perpendicular to the deposit plane in such a way that the cross product H x v = 0 (H and v being the applied field and the velociy of the ions in the electrochemical solution, respectively). In other words, no Lorentz force exists, and thus no surface smoothing effects are expected. In fact, the only gradient effects that are expected should be weak due to the small fieldsthat were applied, (500 Oe maximum).

d) Under these conditions no important effects were expected. We discovered, however, very large effects due to these subtle small fields. We attribute these to the resistance of the deposits without field andwith field. This resistance should be small for applied field due to magnetoresistive effects. When the field is applied the DW's are removed- as already discussed- and the resistance decreases.

In Fig.3a we show experiments revealing SEM micrographs ofthe deposits of Ni on Cu for an anodic potential of -1300mV/SCE with no field. The observed structure here is made of an aggregation of clusters. However, when a field H = 500 Oe is applied in the direction that the ions in the solution move, we observe a arboreous dendritic structure of long branch formations similar to that of sea algae (Fig.3(b)), which is completely different to that of the case when H = 0 (Fig.3a).

This is due to the fact that when the field is 500 Oe the resistance is smaller than tat for zero field applications, because the DW's have been removed. In the zero field case, on the other hand, DW's are present and contribute to the resistance. The effective anodic potential between the end of the filamentous deposits and the calomel reerence electrode is smaller for zero fields due to the larger resistance of the filaments, which does not allow them to grow. Notice that the growth reported here does not have the form of needle-like structures, but has the form of a bead-arboreous-endritic structure; i.e. the filaments have the form of a rosary formed by beads and the filaments do not have any special orientation depending on the applied field. This is at variance with the needle- like structure observed in the deposits of Fe at 200 Oe fields. In addition, the field does not affect the growth of Cu because this is non-magnetic.

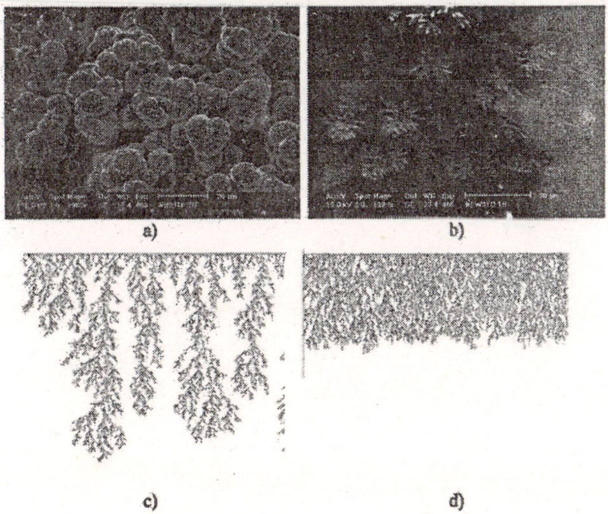

Fig 3 SEM microphotographs of nickel deposits obtained at a cathodic potential of -1300 mV/SCE from a Watt plating solution with the addition of coumarin: a) without and b) with a perpendicularly oriented magnetic field of 500 Oe; and the result of simulation of a cluster growth on the surface for c) k_1=1 and d) k_1=5.

This idea is illustrated by a computer simulation using thc diffusion limited aggregation (DLA) model. Fig.3c is a simulation considering thatthe filaments resistance is zero and then an arboreous dendritic structure is obtained for the deposit in agreement with Fig.3b and the discussion above. However, when the resistance of the wires is taken into account the structure is more compact as descibed in Fig.3d, which correlates well with Fig.3a. These experiments, therefore, show how a subtle magnetic field can provide a gracious and different kind of structure. As in nature, the processes are mostly electrochemical, and it just might be that man natural forms could be influenced by subtleties of the surrounding stray magnetic fields.

Magnetostriction measured with an Atomic Force

One of the problems raised while measuring the variation of the resistance with the applied magnetic field in smallcontacts is: *how much variation of the resistance may be due to change of the contact during the application field because of the accumulated magnetoelastic energy or magnetostriction?* This is a difficult problem to solve and in order to do it we have estblished a new method, that is elegant and educational at the same time, using a local probe; i.e., scanning tunnelling microscopy (STM), atomic force microscopy (AFM), etc. We illustrate it below with the case where we use an AFM.

To observe magnetostricion under the AFM, we place the sample between the poles of a magnet where we can vary the field without moving the magnet itself. An electromagnet is ideal for this application. A schematic diagram of the set up is shown in Fig.4a. While the tip is scannng the surface (typically an area less than a thousandth of a millimetre), the field is activated by turning on the current to a value corresponding to whatever magnetic field strength we desire.

The results are instantaneous, and quite illustrative of th process in action. The sample (typically a wire during our first measurements) was glued to the sample holder on one end, while scanning occurred at the other. In this way we could for example observe the change in the wire's length when thedirection of scan was aligned with both the direction of the magnetic field and the length of the wire. In this case contraction is movement to the left, and expanion to the right.

What is interesting is that depending on the rate of scanning (typically less than a second per line for slow scans), the tip/sample contact is disrupted for a few lines, but then stabilizes immediately after the magnetostriction effect s completed. Even more important, when the field is removed the picture moves precisely back to its original position as if nothing ever happened. We can measure the shift by taking any features immediately before the field is applied and measuring the ditance to the same features after field application. Our results invariably agree with other methods of measurement, yet only this technique provides valuable topographic information of the shift itself on the nanoscopic level.

The method is simple, diret and illustrative in a visual manner of a phenomenon only previously measured indirectly, and its potential is endless with applications ranging from nanotechnology and nanocircuitry to the exploration of new magnetic materials and even with possible extnsions in biology. Besides technological applications the direct observation of magnetostriction holds the potential to expand the understanding of the relationship between mechanical and magnetic properties of materials, and may even lead to the observaton of magneto-mechanical phenomena on the atomic level.

Using this method we have measured the possible displacements or motions of the contacts that give rise to the large BMR depicted in Fig.2b with substantial reproducibility. Within our resolution i the measurement (1 nm), we see no motion whatsoever of the contacts. That implies that if there are displacements they must be much smaller than 1 nm because we do not see any instability whatsoever in the AFM tip when the magnetic field is applied. Therfore, this tends to indicate that the observed values result from magnetoresistance with a high degree of reproducibility. If this is the case, these contact devices clear new ground for very sensitive sensors, especially as reading heads for reading magntic information compacted in the Terabit/inch2. At present research is going on in miniaturization and stabilization over weeks of the electrodeposited contacts.

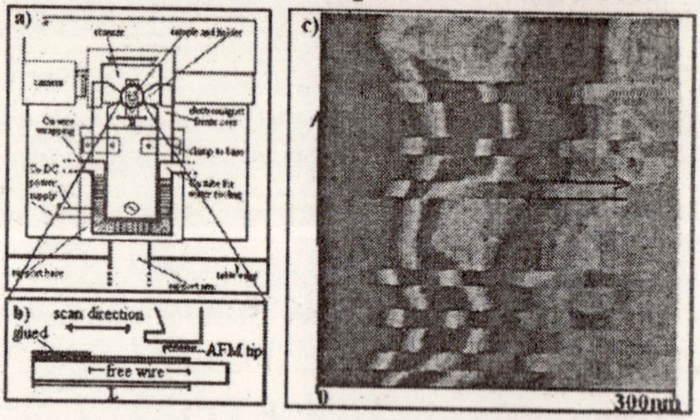

Fig 4 a) Top-view schematic of our setup for the AFM magnetostriction measurement.The sample is depicted in the bold circle between the two poles of the electromagnet and under the scanner.The double arrow marked B indicates the direction of the magnetic field.b) Side-view zoom of the sample configuration, including the AFM tip situated at the bottom of the scanner.The samples were metallic wires, glued on one end, while scanning occurred as close to the free tip of the wire as possible. c) A typical topographic image, of area 300 nm x 300 nm,which results from a tip scan along the length of the wire.The corresponding wire length is from left (toward glued portion) to right on the scan,while the scan itself is moving in the direction of the vertical arrow (bottom to top).The material used is a permalloy (an alloy of nickel and iron), and the length of the wire is about 10 mm. Application of a field causes a linear shift in the whole picture to the left, that indicates contraction, whereas on removal of the field the image moves to the right, and the wire expands back to its original position (see corresponding horizontal arrows).The shift itself was measured at 11 nm for the field strength applied.

THE EXTRAORDINRY PROPERTIES OF MAGNETIC OXIDES

The physical properties of transition metal oxides have been currently the object of many investigatons, since the discovery in 1986, of superconductivity at high temperature in cuprates. In these chemically complex materials, the electronic configuration of the transition elements, varies with its valency and is susceptible to adopt various spin states As a result, strongly correlated electron interactions are generated, leading to complex magnetic and metal-insulator transitions. In this respect, magnetic oxides, containing transition elements such as manganese, cobalt or ruthenium exhibit most fascinting properties.

One of the most famous classes of magnetic oxides concerns the manganites with the perovskite structure (Fig. 1a), which exhibit colossal magnetoresistance (CMR) properties. In these oxides of generic formula $Ln_{1-x}A_xMnO_3$ (A = Ca, Sr, Ba) double exchange phenomena take place, which require a hopping of charge carriers from a Mn^{3+} to a Mn^{4+} ion through an oxygen atom. This effect generates a ferromagnetic metallic state at low temperature, and a metal-insulator transition versus temperatur, which coincides with a ferromagnetic to paramagnetic (or antiferromagnetic) transition.

The latter are also coupled with a structural transition, implying the disappearance of the static Jahn-Teller distortion due to Mn^{3+}, in the ferromagneticstate. As a consequence, at the vicinity of the Curie Temperature T_C a large magnetoresistance can be obtained: the resistance of these materials can be decreased by several orders of magnitude by an external magnetic field of some teslas. A part of thesemanganites, those which contain smaller A cation (Ca^{2+}), exhibit a more complex physical behavior, due to orbital and charge ordering phenomena. In these oxides, the manganese octahedra are occupied by Mn^{3+} and Mn^{4+} species in an ordered way, forming stries of Mn^{3+}distorted octahedra due to Jahn-Teller effect, which alternate with stripes of quasi-regular Mn^{4+} octahedra.

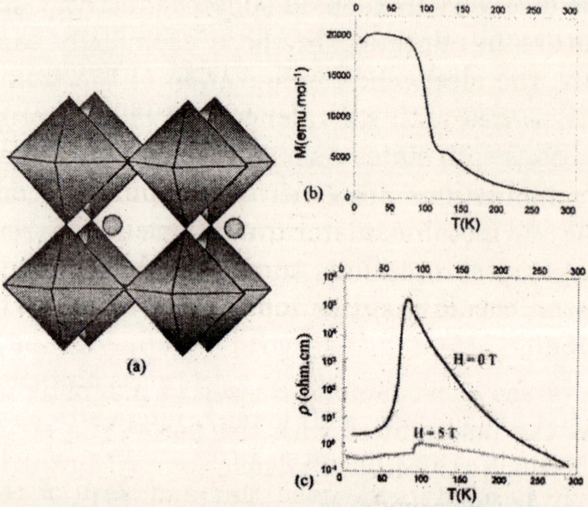

Fig 1 The CMR perovskite manganites $Ln_{1-x}A_xMnO_3$: (a) The perovskite structure built up of corner-sharing MnO_6 octahedra forming cages where Ln^{3+} and A^{2+} cations are located (b) Ferromagnetic to paramagnetic transition, and (c) corresponding insulator to metal transition (H = 0 zero field) and field effect upon resistivity (H = 5T) observed in $Pr_{0.7}Ca_{0.25}Sr_{0.05}MnO_3$.

The physics of the latter compounds is then governed by the competition between ferromagnetism and charge/ orbital ordering, i.e. the orbtal-charge ordered state which is antiferromagnetic and insulating transforms to a metallic ferromagnetic state on the application of a sufficiently high magnetic field, leading also to CMR effect. In these materials, orbital-charge ordering can be destroed by doping the manganese sites with magnetic cations such as chromium or ruthenium, which by magnetic coupling with adjacent manganese induce ferromagnetism and metallicity, so that spectacular modifications of the magnetic phase diagrams of these systes are obtained. In this way, insulator-metal transitions can be induced and CMR effects are enhanced under much lower magnetic fields.

Another interesting property of these manganites deals with their ability to exhibit a spin glass like behavior, for smll A site cations in the composition range close to $CaMnO_3$ (x ~ 0.8 - 0.9). Moreover at the boundary between this region and charge ordered region, CMR properties are also observed.

In fact, the unusual behavior of the manganites is based on a new phenomeon, the electronic phase separation. It has indeed been observed that charge ordered insulating regions coexist at low temperature with small ferromagnetic regions at a submicrometer scale, in a coherent manner in the same matrix. Thus, the CMR effect forthe low T_C manganites results from the percolative conduction through the small ferromagnetic domains embedded in the antiferromagnetic insulating matrix.

Such an electronic phase segregation was also predicted from the large magnetostriction effects obseved in these materials, and described either as dynamic magnetic polarons, or as static phase segregation. Many other properties of these manganese perovskites are closely related to the peculiar double exchange phenomena, such as the photo induced metal-nsulator transition, as well as the electrical field induced magnetic transition that have been recently observed. The curious effect of thermal cycling upon the resistivity, and the step like behavior at low temperature of their magnetization, resistivit and specific heat versus magnetic field at low temperature, are also certainly related to the phase separation that exists in these compounds, and suggest a martensitic type mechanism.

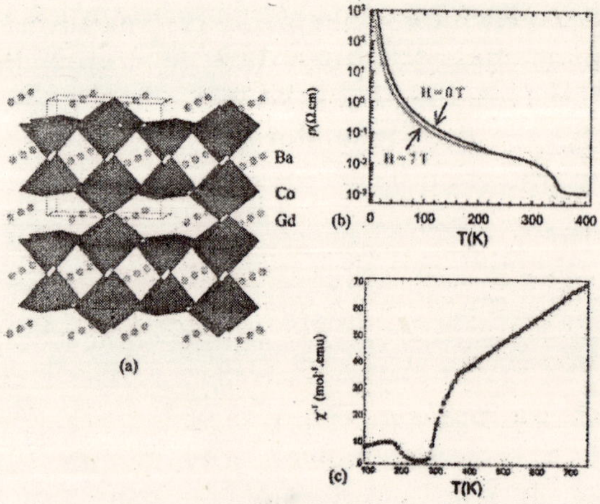

Fig 2 The cobaltites $LnBaCa_2O_{5.5}$: (a) The structure of the ordered oxygen deficient structure built up from rows of CoO_6 octahedra and CoO_5 pyramids, and involving layers of Ln^{3+} and Ba^{2+} cations alternately. (b) The metal to insulator transition of the manganite $GdBaCo_2O_{5.5}$, (c) The corresponding magnetic transition in the paramagnetic regime.

The high degree of spin polarisation of the conduction electrons in hese oxides is of great interest since it can be used to achieve large low field magnetoresistance, which is of capital importance for magnetic recording, or sensing applications. Artificially prepared grain boundaries in thin films allow indeed promising magneto resistance values to be reached. Finally it must be emphasized that the 3D perovskites are not the only manganites which exhibit CMR. High magnetoresistance can also be obtained in the 2D Ruddlesden and Popper manganites, such as $Ln_{2-x}A_{1+x}Mn_2O_7$, bu the low dimensionality of these oxides strongly damages their Curie temperatures. In the same manner, the perovskite structure is not the only one to exhibit magnetoresistance. Large magnetoresistance has also been discovered in the pyrochlore $Tl_2Mn_2O_7$ wich contains only Mn^{4+} species suggesting a different origin of CMR.

Cobalt oxides form also a very important family with extraordinary magnetic properties. This is for instance the case of the compounds $LnBaCo_2O_{5.5}$ (Ln = lanthanide), which exhibit a metl-insulator (MI) transition coupled with a spin transition. These materials are ordered oxygen deficient perovskites, characterized by a layered ordering of the Ln^{3+} and Ba^{2+} cations, so that rows of CoO_6 octahedra alternate with rows of CoO_5 tetragonal pramids (Fig. 2a). As a result, there exist two sorts of trivalent cobalt in the structure at high temperature ($T>T_{MI}$): the octahedral cobalt has a high spin (HS) $t_{2g}{}^4e_g{}^2$ configuration, whereas the pyramidal one has an intermediate spin (IS) $t_{2g}{}^5e_g{}^1$ configration.

In contrast to manganites, the metal to insulator transition appears in these oxides at decreasing temperature (Fig. 2b) around 300 - 350K, and coincides with the transition in the paramagnetic susceptibility (Fig. 2c). This strong coupling betwee magnetism and transport properties is explained by a spin transition of octahedral Co^{3+} species from high spin (HS) $t_{2g}{}^4e_g{}^2$ to low spin (LS) $t_{2g}{}^6$, the $t_{2g}{}^5e_g{}^1$ (IS) pyramidal cobalt being unchanged at the transition. Recent thermoelectric power studies ofthese oxides, corroborate this viewpoint showing that S changes of sign at the transition from n-type in the metallic state to p-type in the insulating state. Moreover, these cobaltites also exhibit negative giant magnetoresistance at lower temperature. Fr instance, resistance ratio higher than 10 can be obtained under 7T (Fig. 2b), in the temperature range T < 200K, where a second transition from the paramagnetic to the antiferromagnetic state has been achieved. Note that negative magnetoresistance has aso been observed in several other cobaltites: the perovskite $La_{1-x}Sr_xCoO_3$ and the bismuth based cobaltites, $(Bi, Cd)_1Sr_2CoO_5$ with a layered structure similar to the 1201-type cuprate.

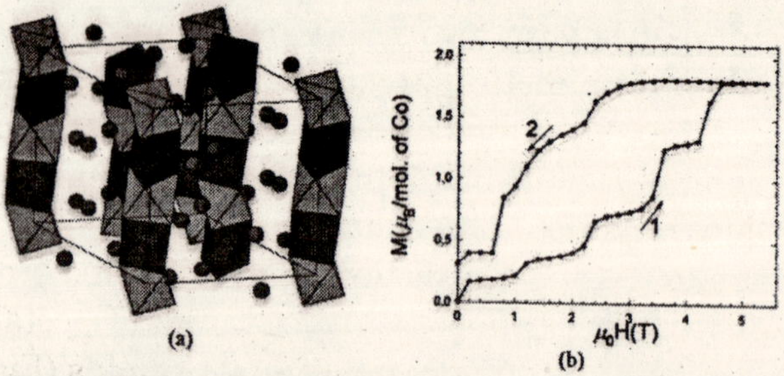

Fig 3 The cobaltite $Ca_3Co_2O_6$: (a) The structure consists of $[Co_2O_6]_\infty$ chains of face sharing CoO_6 octahedra and CoO_6 trigonal prisms forming a triangular array. (b) Magnetization half loop M(H) of this phase registered at T = 2K.

Such a coupling of the magnetic and transport properties in the cobalttes is not limited to the oxygen deficient perovskite structure. A metal-insulator transition correlated with a spin state transition of Co^{3+} is also observed at room temperature in the cobaltite $TlSr_2CoO_5$. In the latter oxide, whose 2D structure is an inergrowth of rock salt type layers "$TlSrO_2$" with single perovskite layers "$SrCoO_3$", the metallic state involves strong ferromagnetic interactions, induced by the cooperative spin transition that takes place at lower temperature.

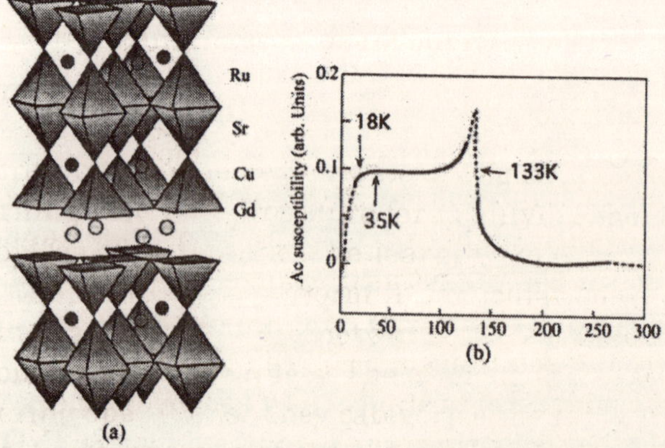

Fig 4 The ferromagnetic superconductor $RuSr_2GdCu2O_8$: (a) The structure consists of single layers of RuO_6 octahedra

sandwiched between layers of CuO_5 pyramids, gadolinium and strontium layers are stacked with a 1-2 order (b) The ac susceptibility curve shows that superconductivity is achieved at 18K (inset at 35K), whereas ferromagnetism appears below 133K according to Chmaissen *et al.*

Another example of cobalt xides is also very attractive for its particular magnetic properties, the one dimensional cobaltite $Ca_3Co_2O_6$. The rhombohedral structure of this phase (Fig. 3a) consists of $[Co_2O_6]$ chains running along the c axis of the hexagonal cell. In each chain, one CoO_6 octahedron alternates with one CoO_6 trigonal prism. The great interest of these compounds deals with the fact that a transition from a ferrimagnetic to a ferromagnetic state is induced on application of a magnetic field. A ferromagnetic intrachain couling exists along $\vec{c}$, whereas an antiferromagnetic intrachain coupling is obtained in the (a, b) plane.

Nevertheless the Co-Co intrachain distances are much larger than the Co-Co interchain distances, so that the magnetism of $Ca_3Co_2O_6$ can be described on he basis of a planar Ising triangular lattice where each chain plays the role of one spin. Thus, the magnetic susceptibility of this phase exhibits two transitions versus temperature at 24K and 12K, which correspond to the setting of the interchain antiferomagnetic coupling and spin freezing (T_f) respectively. In this Ising triangular ferromagnet, AC susceptibility measurements show large shift of T_f from 12 to 16.5K, as the frequency increases by three orders of magnitude. Remarkably, five plateaus can b observed at 2K on the M(H) curve (Fig. 3b), which characteristic magnetic fields are separated by ~ 1.2 T. This feature is reminiscent of the quantum tunneling of mangnetization encountered in high-spin macromolecules.

Ruthenium oxides exhibit also an etremely rich physics from the magnetism viewpoint. Several members of the

Ruddlesden and Popper series $Sr_{n+1}Ru_nO_{3n+1}$, with n = 1, 2, 3, ∞ are metallic ferromagnets with T_C ranging from 105K to 165K, whereas the n = 1 member Sr_2RuO_4, has often been consideed as close to the ferromagnetic order but is in fact a superconductor with a very low critical temperature of ~ 1K. Remarkably, the former compounds, show a Fisher-Langer type anomaly of the conductivity versus temperature at T_C, which is suppressed unde a few T.

In contrast, the calcium homologous phases which are isostructural, exhibit very different properties: the perovskite $CaRuO_3$ (n = 1) is a paramagnetic metal, whereas Ca_2RuO_4 (n = 2) is an antiferromagnetic insulator. As a consequence, solid soluions also show complex magnetic properties, as for example the oxides Ca_2-xSr_xRuO_4, for which a metal-insulator transition, associated with a structural change and magnetic ordering is observed, for low strontium contents. But the most fascinating propertes of ruthenium based oxides have been recently obtained for the ruthenocuprates of the $RuSr_2GdCu2O_8$ family. The physical behavior of this oxide is unique in that, superconductivity and ferromagnetism coexist within the same matrix.

The structure of this hase (Fig. 4a) derives from that of the 92K-superconductor $YBa_2Cu_3O_7$, by replacing the CuO_4 groups by RuO_6 octahedra. It can be described as an oxygen deficient perovskite built up of single octahedral perovskite ruthenium layers sandwiched between pyramial copper layers, layers of Gd^{3+} and Sr^{2+} being stacked according to a "1-2" order, between the CuO_2 and RuO_2 planes. The remarkable coexistence of superconductivity with a T_c ~ 35K, and of ferromagnetism with a T_C ~ 133K is demonstrated from AC susceptiblity measurements (Fig. 4b). Such a behavior shows the important role of the layered architecture of the structure, in order to conciliate these two contradictory properties. In conclusion, these few examples and those well known on iron oxides show thattransition metal oxides represent a vast field of investigation for the discovery of new extraordinary magnetic properties.

OPTICAL STUDIES OF SPIN INJECTION AND DETECTION AT FERROMAGNET/SEMI-CONDUCTOR INTERFACES

Recently, the newly emerging field of spintronics has atracted considerable attention. Significant advances in device performance, in terms of speed, size scaling and power requirements could be achieved by creating spin electronic devices based on the manipulation of spin polarized electrons. Proposed spi analogues to conventional electronic devices have stimulated great interest, *e.g.* the spin polarized field effect transistor (spin FET) and the spin polarized light-emitting diode (spin LED). In order to realize such spin electronic devices, pin dependent electron transport needs to be better understood. It is very important to note that efficient spin dependent transport depends on achieving *both* efficient spin injection from a ferromagnet (FM) into a semiconductor (SC), *and* efficient pin detection of electrons passing from the SC into the FM.

Efficient spin injection has been reported by Fiederling *et al.* and Ohno *et al.* using a magnetic SC but only at low temperature in all SC device structures. Recently, spin injection from FM metal into a SC has been achieved at room temperature with an efficiency of 2 % and 30 % in Schottky barrier structures, and at 80 K with an efficiency of 9 % in FM/AlOx barrier/SC structures, respectively. The question remains as to whethr room temperature efficient operation is possible and also whether strong spin transmission can be achieved between

FM metals and SC. Theoretically, it has been suggested that there may be fundamental obstacles to achieving efficient spin transmission acoss FM metal/SC interfaces via a diffusive electron transport process due to the conductivity mismatch between the FM metal and the SC.

In this case the electron transport properties are dominated by the large resistance of the SC, diluting any spin dpendent effects at the interface (Fig. 1a). It is expected, however, that spin dependent electron transport can be achieved via electron tunneling at FM/SC interfaces. If the tunneling process is spin dependent and the tunnel barrier contact resistane is larger than the resistance of the SC, spin injection and detection efficiencies of up to 100% can be expected (Fig. 1b). So far very few studies have been conducted on spin detection and further clarification of the mechanisms involved is highly desiable.

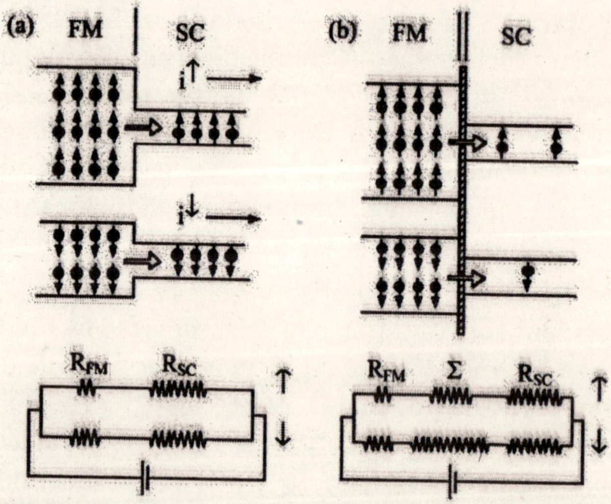

Fig 1 Schematic of spin dependent electron transport at (a) a FM/SC interface and (b) a FM/tunnel barrier/SC interface.A resistor model is also shown for both cases, where R_{FM}, R_{SC} and Σ denote the resistance in the FM and the SC and the tunnel barrier contact resistance, respectively.

We investigated spin filtering across FM/SC Schottky interfaces as a function of FM material, FM layer thickness and applied magnetic field using photoexcitation techniques. Polarized photoexcitation in FM/SC structures was employed to creae a population of spin polarized electrons in the SC substrate (GaAs). The spin dependent electron transport across the FM/SC interface at room temperature was detected as an electrical response, the strength of which varies according to the configurationof the photon helicity with respect to the FM layer magnetization (inset of Fig. 4).

We achieved a change in helicity dependent photocurrent when the magnetization was realigned from perpendicular to parallel to the photon helicity, which is attributed tospin filtering at the FM/SC interface due to the spin split density of states (DOS) in the FM. These spin transport effects increase with increasing FM layer thickness and applied magnetic field. Applying dc measurement techniques, we were furthermore abl to quantify the spin polarization of the electrons filtered at the FM/SC interface. Based on our results we discuss a simple model for the spin transport mechanism across the Schottky barrier.

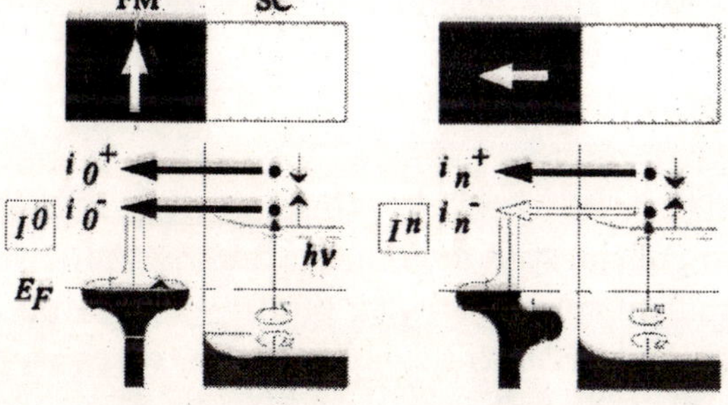

Fig 2 Schematic diagrams illustrating the spin filtering mechanism for electron transport at the FM/Schottky barrier/ SC interface.Averaged density of states of FM is shown for the case of I^0 for simplicity.

We used ultrahigh vacuum (UHV) deposition techniques to fabrcate 2.5, 5.0 and 7.5 nm thick polycrystalline $Ni_{80}Fe_{20}$ and Fe layers directly onto GaAs substrates (Si doped, n=10^{23} m^{-3} and 10^{24} m^{-3} for the NiFe and Fe samples, respectively), capped with 3 nm thick Au layers. An antiferromagnetic Cr sample (5.0 nm, n=10^{23} m^{-3}) was also prepared as a reference. A bias voltage was applied between one Au electrical contact on the surface of the sample and an ohmic contact attached to the back of the substrate. The current flowing through these two pads was measured (both ith and without photoexcitation), while the voltage across the sample was also measured using a separate top contact as shown in the inset of Fig. 4. A circularly polarized laser beam (with photon energy $h\nu$ = 1.96 eV) was used together with an externl magnetic field to investigate the spin dependence of the photoexcited electron current at room temperature. The polarization of the beam was modulated from right to left circular using a photo-elastic modulator with 100 % circular polarization at a freqency of 50 kHz.

As discussed above, the process (tunneling, thermionic emission etc.) by which electrons are transported from the SC into the FM is a key issue for the realization of spin detection. Photoexcitation measurements we recently carried out o band gap engineered FM/AlGaAs tunnel barrier/SC structures clearly showed that significant spin filtering can only be expected for tunneling electrons, in good agreement with the theoretical predictions. Our discussion of the spin dependent tansmission process will therefore focus just on the tunneling of electrons through the Schottky barrier followed by ballistic transport in the FM.

The helicity dependent photocurrent I was measured by modulating the photon helicity from right (σ^+) to let (σ^-). The two helicity values correspond to opposite spin angular momentum values of the incident photons and the helicity gives rise to opposite spin polarizations of electrons photoexcited in the GaAs. The magnetization (*M*) in the FM is aligned prpendicular or in plane (H=0) using an external field. For $\sigma//M$ (or anti-parallel), the electrons in the FM and the SC share the same spin quantization axis, while for σ⊥M, on the other hand, the two possible spin states created by the circularly polarize light are equivalent when projected along the magnetization direction in the FM (Fig. 2).

Consequently, in the remanent state (σ⊥M), since M is orthogonal to the photoexcited spin polarization, both up and down spin polarized electrons in the SC can flowinto the FM, opposing thc electron current from the FM. At perpendicular saturation (σ//M), on the other hand, the up spin electron current from the SC is filtered due to the spin split DOS in the FM. This means that a greater net negative current now flos with σ//M than that for σ⊥M, since the current from the metal to the SC is largely independent of the magnetization configuration. Spin filtering is therefore turned on or off by controlling the relative axes of s and M, and is detected as the helicity ependent photocurrent *I*. With σ⊥M, there is no spin filtering, while spin filtering is turned on by rotating to σ//M.

The helicity dependent photocurrents I^0 and I^n correspond to the magnetization configurations σ⊥M (Fig. 2, left) and σ//M (Fig. 2, right) respectively. I^0 and I^n are proportional to the difference between the current components for right and left circularly polarized light for each magnetization configuration: I^0 μ i_0^+-i_0^- and I^n μ i_n+-i_n-. As shown in Fig. 2, $i_0^+ = i_0^-$ is expected for thecase of the remanent state, while $i_n \neq i_n^-$ is expected for the case of perpendicular saturation. Due to our experimental geometry, where the light enters the SC through the FM layer, we can expect a contribution from magnetic circular dichroism (MCD, i. e. the different absorption of right and left circularly polarized light in a magnetic material) to *I*.

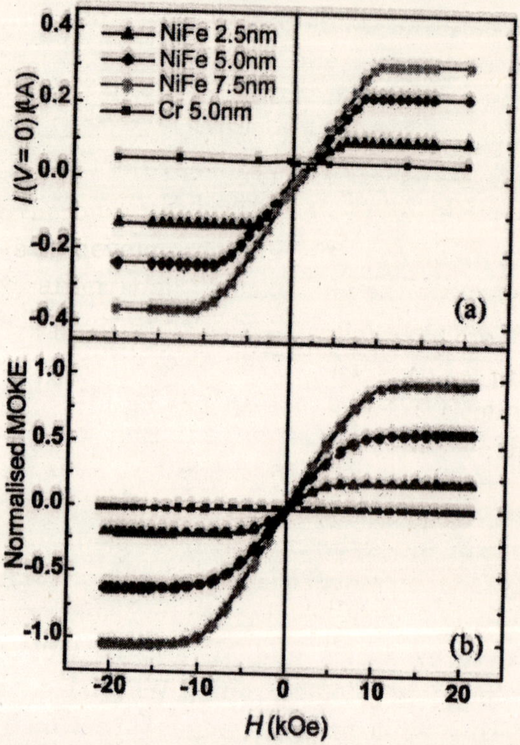

Fig 3 Applied magnetic field dependence of (a) *I* and (b) MOKE signal for NiFe/GaAs (t=2.5, 5.0 and 7.5 nm) and Cr (5.0 nm)/GaAs samples.

The magnetic field dependence of the helicity dependent photocurrent at zero applied bias for the three permalloy samples is shown together with the coresponding polar magneto-optical Kerr effect (MOKE) measurements in Fig. 3 (a) and (b), respectively. The MOKE signal is proportional to the magnetic moment of the FM film and therefore provides a qualitative measure of the FM magnetization.

As can be seenin Fig. 3, the field dependence of the helicity dependent photocurrent matches that of the MOKE signals, showing the magnetic nature of the effect. Similar results were obtained for the Fe samples. We can therefore conclude from this observation that ther are no significant background effects due to Zeeman splitting in the GaAs. Although the Cr sample shows a small offset, the signal does not possess any field-dependence (due to a possible SC-related background), confirming that the Zeeman splitting effec is negligible in our measurement.

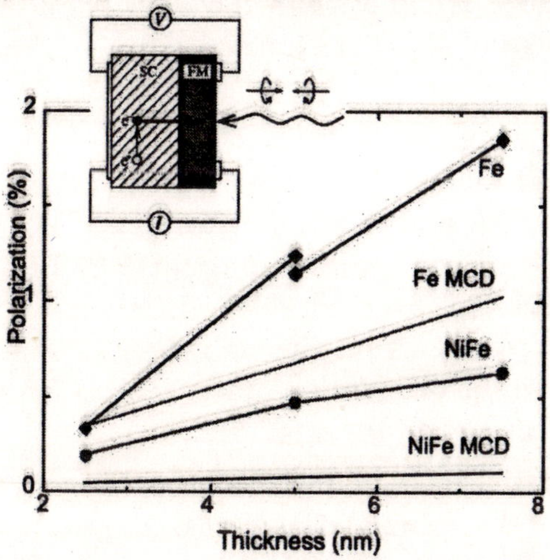

Fig 4 Thickness dependence of spin polarization across the FM/GaAs interface for the case of both NiFe and Fe as the FM.The magnitude of the calculated MCD effects is also shown as positive values. In the inset a schematic of the experimental set up is shown.

In order to study the spin filtering effect quantitatively, we performed dc photoexcitation measurements. Circularly polarized light was generated using a l/4 plate, and the dc helicity dependent photocurrent was observd for both right (I^+) and left circular (I^-) configurations. The spin polarization of the spin filtering effect was estimated as $P = (I^+-I^-)/(I^++I^-)$. Figure 4 shows the thickness dependence of both the estimated spin polarization and the magnitude of theMCD effect for both NiFe and Fe samples.

The spin polarization increases with the FM layer thickness t and is larger than the calculated MCD effect as shown in Fig. 4. A similar thickness dependence of the spin polarization has been reported by van't Erve *et al.*, suggesting that spin filtering occurs in the ballistic regime. It should be noted that the signs of spin polarization for spin filtering are the same for both NiFe and Fe but that the sign of MCD is expected to be opposite for NiFe and Fe.

I conclusion spin polarized electron transport across FM/SC hybrid structures has been investigated for different FM materials (NiFe and Fe), FM layer thicknesses and applied magnetic fields. At room temperature, we observed a clear difference in the helicty dependent photocurrent through the FM/GaAs interface according to the orientation of the sample magnetization with respect to the helicity.

This difference in photocurrent corresponds to a measure of the spin polarized photocurrent passing from the SC nto the FM. The crucial transport mechanism in this spin filtering process is the tunneling of photoexcited electrons through the Schottky barrier. Antiferromagnetic Cr/GaAs shows no spin dependence as expected and provides an important test of the validiy of our experiments.

The spin polarization increases with the FM layer thickness, which provides further support of the view that spin filtering is associated with ballistic transport in the metal. Our combined results unambiguously indicate that spin poarized electrons are transmitted from the SC to the FM with high efficiency.

STRONG ELECTRON CORRELATIONS IN MAGNETIC SYSTEMS

Heavy-Fermion Metals

Strongly correlated electron systems can be well studied with certain rare-earth-or actinide-based intermetallics. At high temperatures, independent local magnetic moments are formed due to he strong repulsive Coulomb force between electrons on the partially filled 4*f*/5*f*-shells of these materials. This local interaction competes with the (upon cooling increasingly strong) coupling between the localized *f*-electrons and the delocalized ligand tates. As a result, the local *f*-moments become progressively screened below the characteristic Kondo-lattice temperature T_K, a process that causes the formation of strongly renormalized ("composite-fermion") quasiparticles, consisting of a local *f*-part an some admixture of itinerant conduction-electron contributions. These "heavy electrons" or "heavy fermions" (HFs) exist in extended regions of parameter space and resemble the conduction electrons of simple metals. However, because of the strong intra-atoic Coulomb correlations, they acquire huge effective carrier masses m* ($\approx$100 - 1000 $m_{electron}$).

Most of the HF metals adopt a symmetry-broken ground state, either a superconducting (SC) or an antiferromagnetically ordered one. In some of the U-bsed HF metals, like UPt_3 and UPd_2Al_3, superconductivity is found to coexist with long-range antiferromagnetic (AF) order (sec.2). On the other hand, for several Ce-based materials superconductivity has been observed to form in the vicinity of an AF instablity, at which the Néel temperature, T_N, vanishes continuously as a function of a control parameter like external or internal pressure.

Exemplary systems are the pressure-induced superconductors $CePd_2Si_2$ and $CeIn_3$ as well as $CeCu_2Si_2$, the first HF metl for which superconductivity at ambient pressure was observed in 1979 (sect. 2). The low-temperature normal (n)-state properties of these latter materials are displaying pronounced non-Fermi-liquid (NFL) phenomena commonly related to critical fluctuaions associated with a quantum phase transition between the antiferromagnetically ordered and a disordered (spin-liquid) state. When applying a sufficiently strong magnetic field, NFL properties usually give way to those of a heavy Landau Fermi liquid(LFL). To study NFL phenomena in more detail, it is highly desirable to search for an AF-QCP in low magnetic fields, i.e., in the absence of HF superconductivity. This appears to be realized in both $CeCu_{5.9}Au_{0.1}$ and $YbRh_2(Si, Ge)_2$ (sec. 3).

Superonductivity

The discovery of superconductivity in the tetragonal compound $CeCu_2Si_2$ containing, at T > TK 15 K, a dense lattice of local magnetic Ce^{3+} moments came as a big surprise, given the antagonistic nature of superconductivity and magnetism. On te other hand, it was found that HF superconductivity is phenomenologically related to the superfluidity of 3He and cannot be explained by the conventional electron-phonon pairing mechanism. In fact, evidence for a magnetic, rather than phononic, Coope pairing was provided ten years later by inelastic neutron scattering (INS) experiments on UPt_3 where AF order with an extremely small moment ($0.02\mu_B$/U) forms below T_N 5 K.

For the hexagonal compound UPd_2Al_3 which enters an antiferromagnetically odered state with a large saturation moment $\mu_s = 0.85\mu_B$/U at $T_N = 14.3$K, HF superconductivity forms and homogeneously coexists with magnetic order below $T_c = 2$K. UPd_2Al_3 may be called a "magnetic HF superconductor", in which magnetism derives from *localize* 5*f* states while Cooper pairs are formed by more hybridized (*delocalized*) 5*f* states, i.e., heavy (composite) fermion quasiparticles. Based upon quasiparticle tunneling and INS work, the acoustic magnon at the center of the Brillouin zone with an ecitation energy ω_E 1 meV could be identified as the "exchange boson" forming Cooper pairs in this strong-coupling HF superconductor.

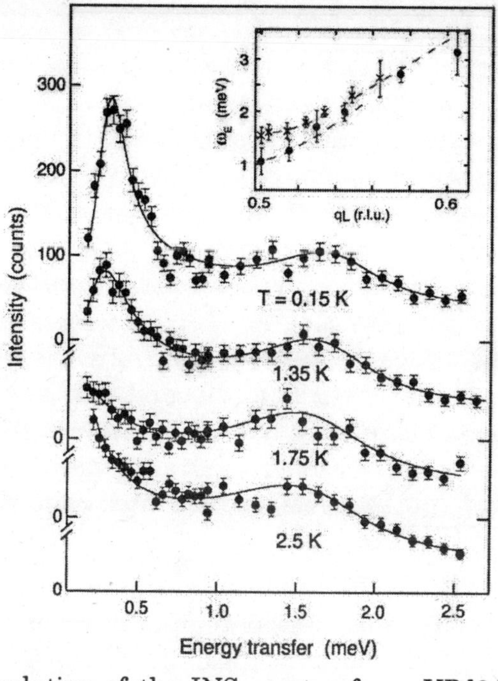

Fig 1 T-evolution of the INS spectra for a UPd2Al3 single crystal at the AF ordering wave vector $\vec{Q}_0$ = (0,0,qL=0.5) through Tc = 1.8 K. Solid lines represent fits using the microscopic "twocomponent" model described in Ref. 9.This procedure may be repeated for other wave vectors $\vec{q}$ close to $\vec{Q}_0$.There is only one fit parameter at given values of T and $\vec{q}$: The magnetic exciton energy.E($\vec{q}$) shown in the inset for T = 2.5 K (crosses) and 0.15 K (circles).The low-energy peak, inelastic at T < Tc and quasielastic at T > Tc, is related to the itinerant HF quasiparticles, the broad hump at higher energies to the magnetic exciton.

This magnon, which, because of the induced singlet-ground-state type of AF order is called "magnetic exciton", may b viewed as a local crystal-field excitation propagating through the U-lattice via inter-site interactions. It thus replaces the optical phonons in classical strong-coupling superconductors like Pb. Thus, UPd_2Al_3 is the first and so far only superconductorfor which a non-phononic pairing mechanism could be demonstrated by the same experimental means as the phononic counterpart in Pb.

The generic phase diagram of $CeCu_2Si_2$ presented in Fig. 2a contains, at small 4f-conduction electron coupling constant g, alow-moment spin-density-wave (SDW) or "A"-phase which coexists in a homogeneous way with weak HF superconductivity below $T_c < T_A$, the magnetic ordering temperature. At moderate coupling, the A-phase is fully replaced by strong HF superconductivity, bt can be recovered if superconductivity is suppressed by applying an overcritical magnetic field. Beyond a critical coupling constant gc, at which TA vanishes continuously characterizing the QCP of this material, pronounced NFL effects were found in the lw-temperature n-state properties of "S-type" single crystals.

Fig. 2b displays the unusual pressure dependence of the superconducting critical temperature of $CeCu_2Si_2$, highlighting a maximum value of 2.3 K at p 2.5 GPa. By deliberately shortening he quasiparticle mean free path through moderate Ge-substitution for Si, Yuan *et al.* could clearly separate two different SC regimes. The low-pressure dome (corresponding to the low-pressure plateau in Fig. 2b) is centered around the AF-QCP, where vitual (high-frequency) fluctuations of the staggered magnetization of the A-phase might be the source to forming massive Cooper pairs. On the other hand, the high-pressure dome (around the maximum T_c in Fig. 2b) appears to coincide with a weak first-oder $Ce^{3+} \rightarrow Ce^{3+}$d valence transition, giving rise to speculations about charge-fluctuation mediated superconductivity.

QUANTUM CRITICALITY

Most of the unconventional properties of the low-T n-state not only of $CeCu_2Si_2$, but also of, e.g., $CeNi_2G_2$ and UBe_{13} can be explained by low-frequency, spatially extended spin-fluctuations which exist in a nearly AF Fermi liquid forming in the vicinity of a 3D-SDW QCP. On the other hand, thorough INS experiments on the quantum critical material$CeCu_{6-x}Au_x$ (x = 0.1) have highlighted the importance of *local* (rather than *extended*) quantum critical fluctuations which are addressed by new theories of a *local-moment* QCP.

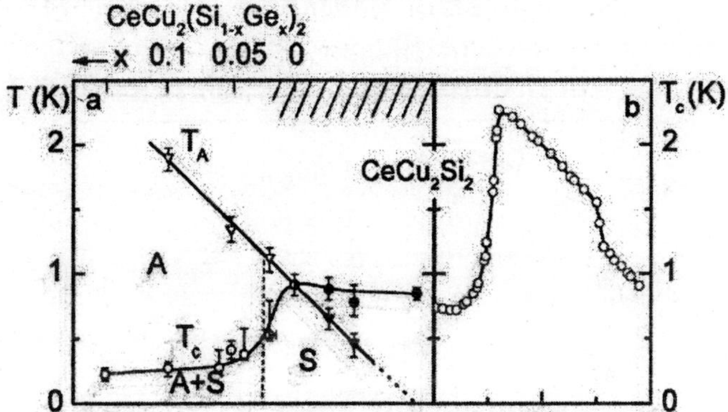

Fig 2 a.Generic phase diagram of $CeCu_2Si_2$ combining data obtained with undoped polycrystals from the homogeneity range of the ternary chemical phase diagram (hatched) and from Gedoped ones, $CeCu_2(Si_{1-x}Ge_x)_2$. Since the temperature T_A of the SDW transition increases linearly with the Ge-concentration x, the 4fconduction electron coupling constant g was assumed to be linear in (1-x), i.e., $(1-T_A)$. b. Pressure dependence of T_c up to p = 8.5 GPa for $CeCu_2Si_2$, see Ref. 12.The different Tc values at p = 0 in a and b relate to two polycrystalline samples with slightly different stoichiometry (Cu content).

Recently, Zhu *et al.* proposed that measurements of the Grüneisen raio $\Gamma \sim \beta/C$ (where β and C are the volume thermal expansion and specific heat, respectively) may well be suited to distinguish between different QCP scenarios. As these authors argue, G(T) must diverge at any QCP. This has indeed been observed fo both paramagnetic $CeNi_2Ge_2$ (inset of Fig. 3a) and the isostructural weakly antiferromagnetically ordered compound $YbRh_2Si_2$ (T_N = 70 mK). In the main part of Fig. 3a the critical Grüneisen ratio, $\Gamma^{cr}(T)$, as given by the singular parts of both b(T) and C(T, is shown to diverge ~ T^{-x} with x = 1 for $CeNi_2Ge_2$. This agrees well with the prediction for the 3D-SDW scenario. By contrast, recent investigations of $YbRh_2Si_2$ revealed $\Gamma^{cr}(T) \sim T^{-x}$ with x = 0.7 as expected in the presence of *local* AF quantum ritical fluctuations (Fig. 3b).

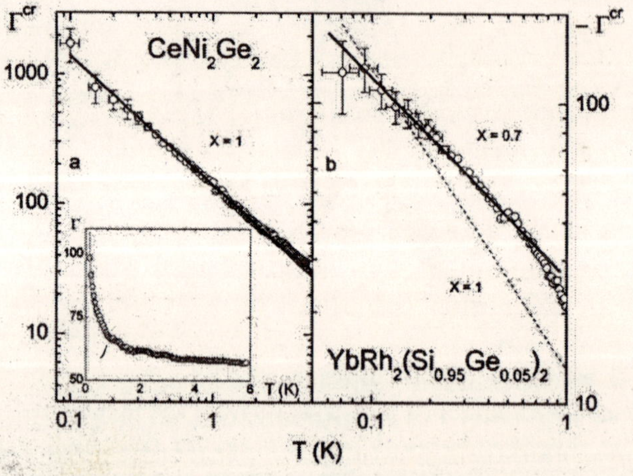

Fig 3 Critical Grüneisen ratio $G^{cr} = (V_m/k_T)\ (B^{cr}/C^{cr})$, as $(-)\Gamma^{cr}$ vs T in double-logarithmic plots for $CeNi_2Ge_2$ (a) and $YbRh_2(Si_{0.95}Ge_{0.05})_2$ (b) [V_m:molar volume;.T: isothermal compressibility; $\beta^{cr} = \beta(T)$-bT and $C^{cr} = C(T) - (\alpha T + \delta/T2)$ are the critical components to the volume thermal expansion and specific heat, respectively, cf. Ref. 18]. Inset shows $\Gamma(T) \sim \beta(T)/C(T)$ as measured (on linear scales) for $CeNi_2Ge_2$.

A QCP at zero can, in principle, be approached by moderate volume expansion of $YbRh_2Si_2$. In fact, in the case of a slightly doped $YbRh_2(Si_{1-x}Ge_x)_2$ single crystal with a nominal Ge-concentration x = 0.05, the Néel temperatre was found to be as low as T_N = 20 mK. By tuning the system away from its QCP, i.e., through applying a magnetic field (^c) larger than B_c 0.027 T necessary to suppress the weak AF order in this Ge-doped crystal, the field dependences of both theelectronic specific heat, $C_{el}(T,B) = g_0(B)T$, and the electrical resistivity, $Dr(T, B) = A(B)T^2$, could be attributed to the field-induced low-T LFL state.

From the divergences of both the Sommerfeld coefficient $g_0(B)$ and the resistivity coefficient A(B) asB B_c^+, as well as from those of $C_{el}(T)/T$ and $A(T) = Dr(T)/T^2$ as T 0 at $B = B_c$ it was concluded that field (temperature) acts as the only relevant energy scale at low temperature (field). Two striking observations made with such $YbRh_2(Si, Ge)_2$ sinle crystals are worth mentioning:

(i) Correlated, but unscreened large paramagnetic Yb^{3+} moments are observed in the static bulk susceptibility (inset of Fig. 4a) as well as in the low-temperature electron-spin resonance.

(ii) The temperature dependnces of $\gamma_0(T) = C_{el}(T)/T$ and $\Delta\rho(T)$ taken at B $B_c(0)$ behave *disparately*: $\gamma_0(T) \sim$ (- logT) and $\Delta\rho \sim T$ are observed upon cooling from T 10 K down to T 0.3 K. While the linear T-dependence of the resistivity can be followed all the way down to 10 mK, th lowest temperature of the experiment (Fig. 4b), $\gamma(T)$ at T = 0.3 K is found to diverge stronger than logarithmically, i.e., $\sim T^{-\varepsilon}$, $\varepsilon = 0.4 \pm 0.03$ (Fig. 4a). This disparity hints at a break-up of the composite fermions on the approach to the QCP.

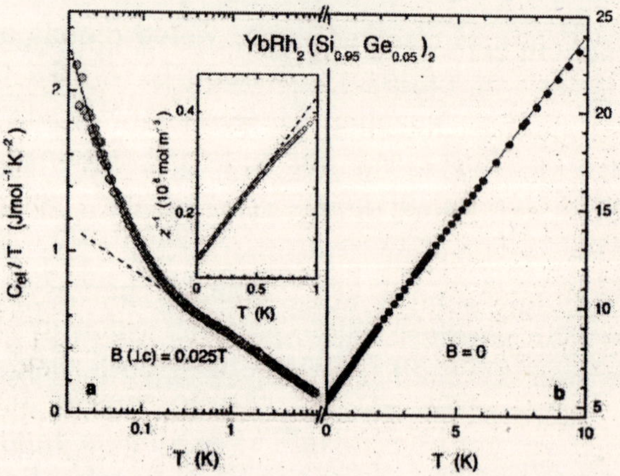

Fig 4 Electronic specific heat divided by T as a function of T (on a logarithmic scale) measured at a field (⊥c) of 0.025 T, close to $B_c(0) = 0.027$ T (a) and electrical resistivity. vs T (on a linear scale) measured at B = 0 (b) for a $YbRh_2(Si_{0.95}Ge_{0.05})_2$ single crystal. Dashed [solid] line in a represents $C_{el}/T \sim$ (-logT) [$\sim T^{-(0.4\pm0.03)}$], solid line in b represents $\rho-\rho_0 = aT$ with ρ_0 5 μΩcm. Inset displays the zero-field ac-susceptibility as χ^{-1} vs T for the same crystal. In the same T-range (T < 0.3 K) where C_{el} (T)/ T deviates from –logT, χ(T) follows a Curie-Weiss law, $\chi^{-1}\sim$(T - Θ) with Θ-0.3 K. From the slope a large paramagnetic moment, μ_{eff} 1.4 μB/Yb^{3+}, is derived.

Strong electronic correlations on partially filled localized 4*f* or 5*f* shells, weakly hybridized with the itinerant ligand states, cause the formation of extremely heavy quasiparticles composed of a local spin part and delocalized charge-carrier contributons.

They may form Cooper pairs, the SC glue presumably being of magnetic origin, at least in most cases. This could be convincingly demonstrated with the aid of tunneling and neutron scattering experiments on UPd_2Al_2.

In case of a number of Ce-based HF etals, superconductivity was found to be intimately related to the existence of an AF instability, most likely of the 3D-

SDW type. On the other hand, the QCP in slightly Ge-doped $YbRh_2Si_2$ appears to be of the *local-moment* variety, although alread in undoped $YbRh_2Si_2$ the ordered moments below T_N = 70 mK are extremely small, in contrast to the large paramagnetic moments, cf. inset of Fig. 4a.

Interestingly enough, superconductivity could not be observed in this material. To find out whether a *local-moment* QCP is indeed unfavorable for HF superconductivity has to be left for future work.

10

Nano-magnetic Structures Characterization in Thin Films

The driving force for developments in nano-technology results from the increasing demand related to key technologies like microelectronics and nano magnetic devices. In decreasing scale of many devices, high-resolution characterization methods have become of fundamental importance for further development in nano technologies. The availability of powerful microscopy methods, is certainly important for the development of new and functional materials. The characterization of materials at various and increasing levels of resolution, structures, microstructure, and defect geometry, as well as chemical composition and spatial distribution are important parameters determining the behavior of materials and practical applications.

INTRODUCTION

Over the past several decades, amorphous and more recently nano-crystalline materials have been investigated for applications in magnetic devices. One example is the high density magnetic random access memory technology which has grown over the past decade due to its potential to store more data, access that data faster and also to use less power than current memory technologies (however, the reliable fabrication of the notches in small (100 nm) rings could present a technological challenge). Demands for the continuous increase in the data storage density bring the challenge to overcome physical limits for currently used magnetic recording media.

Benefits was found in the nano-crystalline alloys because of their chemical and structural variations on a nano-scale which are important for developing optimal magnetic devices with high properties. Magnetic nanoparticles or nanowires are attracting much attention because they offer the opportunity to study magnetism in between the atomic and bulk limits and because ordered arrays of ferromagnetic particles or wires are of potential interest for applications such as ultra-high-density magnetic recording devices. In the last decade, the introduction of new experimental techniques, such as electron holography, super conducting quantum interference device magnetometry (SQUID) and magnetic force microscopy, has provided direct ways to probe the magnetic properties of individual nano particles. The aim of this paper is mainly to point out some methods of structural characterizing tools like:

1. Atomic force microscopy (AFM)
2. Magnetic force microscopy (MFM)
3. Scanning tunneling microscopy (STM)
4. Lorentz transmission electron microscopy (LTEM)

Fundamental calculations with sensitive models to identify the properties of the nano-structures are too important.

MAGNETIC FORCE MICROSCOPY

The magnetization reversal of two-dimensional arrays of parallel ferromagnetic Fe nanowires embedded in nanoporous alumina templates has been studied. The application of field-dependent MFM provides a microscopic method to obtain the hysteresis curve of the array, by simply registering the fraction of up and down magnetized wires as a function of applied field.

The observed deviations from the rectangular shape of the macroscopic hysteresis loop of the array can be ascribed to the spatial variation of the dipolar field through the inhomogeneously filled membrane. The system studied proves to be an excellent example of the two-dimensional classical Preisach model, well known from the field of hysteresis modeling and micromagnetism.

FABRICATION AND EXPERIMENT

A standard 4 inch silicon wafer (100) with a 100 nm thick, thermally grown SiO_2 insulating layer was spin coated with 0.4 μm double-layer resist and baked. The substrate was than placed in an e-beam writer and the desired pattern was written. Using e-beam evaporation in a vacuum chamber with a typical evaporation rate of 0.1A, a 25 nm permalloy film was deposited onto the holes in photoresist. After lift-off in acetone the array of rings was obtained. MFM measurements were done using lift mode magnetic force microscopy.

One may distinguish a large bundle of wires originating from a region of filled pores with highly uniform, straight, and parallel Fe nanowires

STABILITY MEASUREMENT OF PATTERNED MEDIA USING MFM

The thermal stability of the patterned media was measured using MFM. The sample was saturated in a field of 20 kOe, and MFM images of the patterned region were acquired after aging the sample at room temperature in zero applied field for times up to 1.5×106s.

The equation which explains the relation between relaxation time of an information on to a magnetic particle (the time constant t for decay over an energy barrier "E or "E mag is taken to follow the Arrhenius-Nel laws) as the following, E cal = K BT(1) Progress In Electromagnetics Research Symposium 2005, Hangzhou, China, August 22-26 671 Hence, t = 10 "10exp "E mag E cal (2)

Standardmagnetic force etched silicon probes were used for the measurement. The lift height used was 50 nm. The samples containing the rings were magnetized using two SmCo block magnets. The field between the two magnets was tabulated as a function of the distance of separation between them.

NANO METER RESOLUTION TECHNIQUES

Magnetic force microscopy is one of tools for investigating magnetism on a small scale. Techniques such as alternating gradient magnetometry (AGM), vibration sample magnetometry (VSM), SQUID and imaging via the magneto-optical Kerr effect (MOKE) allow for the collective magnetic behaviors of arrays of nanomagnets to be characterized. However, these techniques lack nano-meter resolution and thus are not ideal when studying individual nano-magnets. To study individual elements of an array there exist several techniques with spatial resolution capable of characterizing individual nano-magnets. These techniques include magnetic force microscopy (MFM), Lorentz electron microscopy (LEM), scanning electron microscopy with polarization analysis (SEMPA), spin polarized scanning tunneling microscopy (SPSTM), spin polarized low energy electron microscopy (SPLEEM) and MOKE microscopy.

To fully characterize nanomagnets and ensembles of nanomagnets, a combination of these techniques should be used. The principle behind LEM is the Lorentz force which describes the motion of an electron through a magnetic field, classically given by: F = qvB (3) Instrument like Nano " R TM is a general purpose atomic force microscope (AFM) for making routine images on structures with nanometer sized features. The three primary subsystems a control unit, AFM are the master computer, and the Nano " R TM stage.

CONCLUSION

Magnetic nano structures are subjects of growing interest because of their potential applications in high density magnetic recording media and their original magnetic properties. Investigations on nanomagnetic structures becoming very important for designing modern magnetic devices that contains these nano-magnetic shapes and structures. It must be taken in count parallel with the surface, shape, and size properties identification (through correct microscopy) before the manufacturing process of the nano-structured devices begins By combining bulk magnetization measurements (superconducting quantum interference device magnetometry) with field-dependent MFM, it was possible to decompose the macroscopic hysteresis loop in terms of the irreversible magnetic responses of individual nanowires. The latter are found to behave as monodomain ferromagnetic needles, with hysteresis loops displaced asymmetric! as a consequence of the strong dipolar interactions between them.

MAGNETIC RECORDING MEDIA FOR EXTREMELY HIGH DENSITY DATA STORAGE

Magnetic recording media is one of most important components in any magnetic recording system. The state-of-the-art and research into hard disk media is approaching nano-scale and sub-nanoscale engineering in the lateral and vertical directions. Nanoscale means that the dimensions smaller than 1 billionth of a metre. In magnetic media, information is stored in a group of tiny magnets.

In a digital recording scheme, if similar poles of two groups of magnets face each other, then it is considered a 1, where as if opposite poles face each other, it is considered a 0.

Recently we achieved small grains with size below 6 nm, which will enable a 2.5 inch diameter hard disk to have 300 trillion small magnets. However, when the grain size is reduced down to this scale, a phenomenon called super-paramagnetism poses a limit on the magnetic properties. This is one of the key challenges for longitudinal recording media.

To overcome super-paramagnetism, Laminated Antiferromagnetically Coupled (LAC) media depicted in Figure 1 can be used. LAC media has at least two magnetic layers with moments pointing in opposite directions.

The principle of LAC media is based on adding extra energy in the form of antiferromagnetic coupling to stabilize the bits. To make LAC media, one needs to deposit just two more thin layers (a magnetic layer and a ruthenium spacer layer of 6 Å thickness) in addition to the existing layers in conventional media (Figure 1).

The ruthenium layer has to be deposited with a precision of 0.1 nm over the whole disk. With LAC media, the thermal stability of the written bits improves dramatically, from a storage life of a few months to few tens of years. It is believed that the LAC media can take the recording densities to as high as 300 Gb/in2, which means we can reach hard disks with a capacity more than 600 GB.

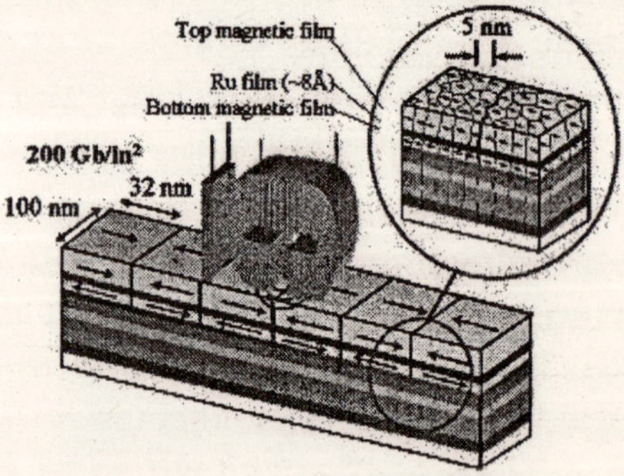

Figure 1: Schematic illustration of laminated antiferromagnetically coupled (LAC) media and the magnetic recording process.

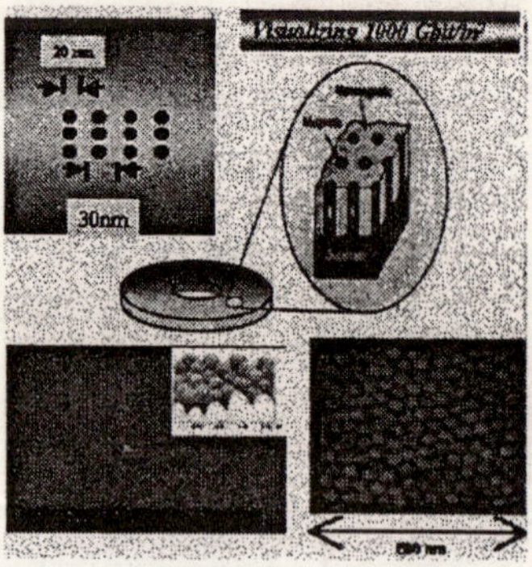

Figure 2: Schematic of patterned media and the patterned structure obtained by ion beam bombardment self-assembly.

There are two types of magnetic recording modes: longitudinal and perpendicular magnetic recording. In perpendicular recording, the bits are stored by arranging magnets vertically, with opposite poles facing each other and is therefore more stable at high-storage densities. It is believed that perpendicular recording technology will take over the existing longitudinal technology in the near future. We are working on various kinds of perpendicular recording media such as Co/Pd multilayers and FePt films. In a media based on Co/Pd multilayers, there are about 10-30 layers of Co (3 Å) and Pd (6 – 8 Å). These materials have potential to support densities up to 1 trillion bits per square inch.

In both the longitudinal and perpendicular recording, one bit is stored in a group of many small grains and is therefore thermally unstable. However, in patterned media recording, one bit is stored by one grain and therefore, the grain can be larger (Figure 2). Therefore, the integrity of the data can be maintained even at densities higher than 1 trillion bits per square inch. In patterned recording media technology, producing the media at a cheaper cost is the key issue and we are carrying out research on various ways of producing patterned nano-structures. One of the ways is to self-assemble the nano-structures by chemical routes or ion-beam bombardment. Other ways to produce nano-structures are by electron beam lithography and nano-imprinting technology. We have patented a new technology based on radiation induced nano-scale phase change to produce magnetic nano-structures. We are now combining the nanoself-assembly method with nano-imprinting technology to produce patterned media with long-range order on large area.

11

Magnetic QCA Systems

INTRODUCTION

Quantum-dot cellular automata (QCA) is deemed a potential candidate to replace or supplement CMOS electronic circuits at the end of the 'roadmap'. QCA effects computation through the interaction of fields, and requires the transfer of charge only within a small, confined volume, called a QCA cell. In most cases the fields are electrical in nature, and electrons move within a cell in response to the charge positions of neighboring cells. It has been shown theoretically that an array of cells can be used to effect complex computations, similar to that of a CMOS processor.

The QCA concept is independent of the material system in which it is embodied. Demonstrations of QCA logic gates, latches, and power gain using metal tunnel junction technology at cryogenic temperatures have been performed. Gardelis et al. have demonstrated coupled semiconductor quantum dots toward QCA. Lent proposed using self assembled molecular layers for supporting charge containment, and progress toward developing the technology for building these structures is in progress. DNA has been proposed by Sarveswaran et al. as an alternative material for templating the molecular charge containers. Another promising materials system, i.e. that of nano- magnetics, is the subject of this paper. The advantages of using magnetic materials to transfer information by magnetic dipoles rather than electric dipoles, as in the material systems mentioned above, include the simplicity of fabrication and robustness, for example, true room temperature operation. Additionally, magnetic materials are insensitive to radiation, suggesting applications of magnetic QCA requiring robust performance and low power consumption in harsh operating environments such as space, satellite and military applications. This paper reviews progress towards developing networks of nanomagnets suited to QCA operation, and points towards the realization of all magnetic systems, including logic, memory, and power gain.

BEHAVIOR OF NANOMAGNETS

As is well known, macroscale ferromagnets are composed of numerous magnetic domains, within each of which the magnetic dipoles are aligned due to quantum-mechanical exchange interactions. In the presence of a sufficiently large external field, the magnetic domains align, leading ultimately to a saturated magnetization state of the magnet.

When the external field is removed, most of the domains remain aligned, and a remanent magnetization remains. This effect underlies the familiar hysteresis of M–H (or B–H) magnetization curves, where H is magnetic field intensity, M is magnetization of the magnet, and B is the magnetic flux Microelectronics density. Upon driving the magnet in the opposite direction, the domains reverse direction and the magnetization is reversed. At the point of zero magnetization, where no domain direction is preferred (i.e. zero crossing of the M–H curve), we define the value of the external H-field as the coercive field, or the coercivity. A magnet with high coercivity is said to be 'hard' and one with low coercivity, 'soft'. The value of coercivity in nanomagnets depends on their dimensions in the direction of the applied field, as well as their aspect ratio. This shape anisotropy will be shown to be critical in the application of nanomagnetics to QCA.

For simplicity, we can define a nanomagnet as one whose size is sufficiently small that the domains are relatively few, so that the behavior of the domains is somewhat predictable, controllable, and shape dependent. Fig. 1 shows the average intrinsic coercive field as a function of particle size of a collection of spherical nanomagnets. For nonspherical magnets, above a critical average size, Ds depends on material, shape anisotropy, and crystal anisotropy, and the particles are multidomain. That is not to say that they are always in a multidomain configuration, but that they can be so. At and below Ds, the particles are on average single-domain and exhibit the maximum intrinsic coercivity. As the particles shrink, thermal energy, proportional to the particle's volume, competes with the magnetostatic energy, and the coercivity decreases. At some point, Dsp, thermal effects win out, the domains cannot be maintained, and the particles become superparamagnetic.

For a single nanomagnet, the M–H curve exhibits discrete jumps rather than smooth behavior due to the small number of individual domains. Hence, the coercivity of a single particle whose size can be imagined to shrink would pass through the few-domain regime in which domain walls enable switching to occur more easily through the flipping of relatively few magnetic dipoles, to single-domain, highest coercivity state, in which no domain wall exists to nucleate switching, so that all the dipoles must flip together. Such switching of these domains can take place on time scales of a ns, which can be considered to be the upper limit to the speed of magnetic QCA.

Our dots, which range in dimensions up to about 300 nm, are somewhat larger than Ds (typically around 100 nm), but are still on the slope of the curve near Ds. By using non-spherical magnets, we can tailor the aspect ratio to provide for us a hard, i.e. short, and easy, i.e. long, axis, which allows us to define directions for coupling in the presence of an external field. As the aspect ratio of a nanomagnet increases, there is the further effect of additive magnetostatic energy and coupling along the length, so the coercivity is further increased. Hence, for a given width of magnet above Ds, longer magnets have more of a tendency to be single domain while shorter ones tend to break up into multiple domains. The following section discusses how elongated nanomagnets can be arranged to effect QCA behavior.

MAGNETIC COUPLING AND NANOMAGNETIC QCA

Most magnetic thin films, such as permalloy or cobalt, display in-plane magnetization. That is, the preferred domain orientation is parallel to the plane of the film. Other films, such as carefully constructed CoPt multilayers, can exhibit out-of-plane magnetization. Coupling of magnets in either of these configurations can, in principle, be used for QCA. Fig. 2(a) and (b) show how two magnets can be assigned logic states of '1' or '0', and can couple either in a ground state, or higher-energy metastable state. It turns out that the typical energy difference between the two states is on the order of 100 room- temperature kT, or more, depending on the shape, size and coupling, suggesting the use of nanomagnet QCA at elevated temperatures. It is not hard to see how a chain of narrowly. Clocking of magnetic QCA devices. (a) Adiabatic clocking of QCA wire, and (b) majority gate with high coercivity input magnets spaced nanomagnets (Fig. 2(c)) could be used as a QCA 'wire', as first investigated experimentally by Cowburn and Welland and Cowburn. Basic QCA theory involves adiabatic switching through a null state to minimize dissipation. To achieve this in electronic configurations, the electrons are first allowed out of their dots within the cells by lowering the energy of their charge container walls (the manner of which depends on the chosen material system), the inputs are then applied, and finally the walls are raised adiabatically.

In so doing, the array of cells remains near the ground state until the walls are fully raised and the electrons have taken on the ground state configuration with no metastable regions, or 'kinks', along the line.

The clocking field, H ext, is applied in the direction of the hard axis. As H ext increases past the intrinsic coercivity of the wire components, their magnetization is forced in the same direction. However, due to shape anisotropy, the input cell having higher intrinsic coercivity retains its initial value. Power gain occurs by the energy transfer from the external clock field to the high internal energy state of the magnets (middle wire of Fig. 3(a)). This energy goes into switching as the clocking field is relaxed.

As the clock field is adiabatically lowered, the wire relaxes into the ground state with no kinks. This scheme can be extended to a majority gate, as shown in Fig. 3(b). Here, three input magnets of higher coercivity couple toward a center magnet whose magnetization is the majority vote of the three inputs, and which then is launched down the wire as in Fig. 3(a).

One feature of this scheme is that it requires out-of-plane magnetization. This requires a plating process for fabrication, since conventional lift-off tends to leave conical-shaped structures as the opening in the resist at the surface closes off from the metalization. Other out-of-plane structures could be fabricated from machining multilayers of CoPt that exhibit proper magnetization direction, and such work is now in progress.

In-plane magnetization can also be used for defining QCA logic gates. We have performed extensive experiments defining variously shaped nanomagnets in order to optimize their coupling. Fig. 4 shows various shapes of permalloy films fabricated by electron beam lithography and liftoff.

The MFM image of a single domain magnet looks like a closely coupled dark and light spot. Each pair of nanomagnets that are antiferromagnetically (AF) coupled (anti-aligned) shows light and dark spots on diagonals. Fig. 5(a) shows two trials of six different shapes, with a total of 144 pairs per shape divided among two different aspect ratios and three different sizes.

The shapes are shown to the right of the MFM images. It can be seen that the larger sizes toward the top of each section are all AF coupled, while as they get smaller they are ferromagnetically coupled (aligned), as evidenced by the horizontal light and dark streaks.

Careful inspection reveals that the locations of the smallest pairs that do not exhibit the desired AF coupling are mostly consistent between the two trials, suggesting that their aberrant behavior is due to defects in the fabrication process. Fig. 5(b) shows the results of a statistical analysis of errors in QCA wires for various shapes shown in Fig. 4.

The numbers represent the average ordering lengths for the different shapes, determined from the average of three independent demagnetization processes of nanomagnet chains in different sizes and aspect ratios. The average ordering length of a single chain is approximated by the length of the chain divided by the number of parallel- aligned dipole pairs in the given chain.

Fig. 6 shows the AFM and MFM images of a line of 64 rectangular magnets exhibiting perfect AF behavior down the entire line. Such strong coupling is totally outside the realm of random chance, and suggests that using nano- magnets for QCA coupling is indeed feasible. Fig. 7 shows one possible embodiment of in-plane, trapezoidal nano- magnets configured as a majority gate, as simulated by OOMMF.

(Parish and Forshaw have modeled a similar configuration.) The three horizontal (relative to the figure) trapezoids to the left are the inputs that can be set and held at their logical values by current carrying wires during the clocking phase.

As the horizontal clocking field increases, the vertical magnets are magnetized horizontally, and when the clocking field decreases, they are allowed to relax to the ground state, vertically magnetized configur- ation. In the figure, the input magnets are all magnetized to the right, which in this case signifies a '1', for the top two magnets (because of their relative configurations with the vertical direction of the coupling magnets) and a logic '0' for the bottom dot. As expected, the central dot is influenced by the majority vote and is magnetized up, starting the signal propagating down the wire.

SUMMARY AND CONCLUSIONS

We have discussed some of the interesting behavior of magnets on the nanoscale and related it to applications for quantum-dot cellular automata. We presented the utility of shape anisotropy for various configurations of nanomagnet QCA. We showed that the reliability of coupling is shape dependent, but have not yet determined the optimum shape.

Simple rectangular magnets were shown in one case to couple with no errors between 64 stages. With more effort, we believe that the advantages of nanomagnet QCA, i.e. ease of fabrication, robustness at room temperature, nonvolatility, straightforward I/O, and potential integration with other magnetic systems such as magnetic random access memory (MRAM), make this a potentially useful technology in spite of its relatively low speed compared with conventional electronic devices.

12

Spin Electronics and Nanomagnetism

BACKGROUND TO THE NETWORK AND TECHNOLOGY STATUS

Spin dependent transport was discovered in Europe in 1988, by our co-workers A. Fert and P. Grünberg, with the so-called Giant Magnetoresistance Effect (GMR) in metallic multilayers. Its potential was such that it took less than 10 years from discovery to the >1 billion $/year market of magnetic recording. European laboratories have since maintained a world level of expertise in all sub areas of Spin Electronics such as magnetic multilayers, magnetic semiconductors, magnetic nanostructures, spin dynamics, or spin injection in semiconductors.

The next step came in 1995 with the introduction of the Tunneling Magnetoresistance (TMR), which provides a more scalable configuration and a much higher magnetoresistance effect. This first heterostructure (metal+insulator) resulted in the magnetic random access memory (M-RAM) which is about to impact the semiconductor industry (product in 2004). A wealth of magnetoresistive sensors is being developed in parallel for applications in automotive industry, biomagnetic tagging, banking (cards, bank notes) etc. Now the field is bursting with new phenomena such as 100% spin-polarized half-metallic ferromagnets, spin injection into semiconductors, magnetic semiconductors, current control of the magnetization switching, etc., opening the way to many novel *hybrid devices* such as for instance spin transistors with gain.

Such devices will have to keep pace with today's Moore's law driven evolution of size as well as computing and telecommunication bandwidths. Confinement-related (mesoscopic, quantum) effects and processing speed will become critical, and thus fundamental understanding of the underlying phenomena such as electron transport, and also new materials, will be the key for the sustainable development rate of these devices. Spin Electronics and Nanomagnetism are thus inseparable because the amplitude, reproducibility, reliability and time response of devices are generally driven by the magnetic configuration or by the spin coherence within a nanostructure. This becomes crucial as dimensions approach the fundamental length scales of magnetism (<10 nm).

AIMS AND OBJECTIVES OF SPIN-NET

The proposed Network of Excellence is motivated by the vision shared by all our partners that *Spin Electronics is becoming a new paradigm in nano-electronics: from charge to spin of the electron, with the prospect of higher speed, lower power consumption, better scalability and non-volatility, furthermore offering a way to quantum information processing.* This paradigm will be at the heart of tomorrow's magnetic, electronic and photonic materials & systems that will drive the sustained technology revolution towards a more efficient and safer life within Europe as well as a stronger industrial competitiveness.

The main objective of our Network SPIN-NET is thus to advance fundamental and basic science and training to establish Europe as the world scientific leader and stronghold on Spin Electronics, by integrating within a single Network of Excellence the research activities of the best European teams working on all main aspects of the field.

NEED AND RELEVANCE

The domain of "Spin Electronics and Nanomagnetism" strongly contributes to at least two priority areas of FP 6. The more basic approach is materials oriented and relates to Thematic Area 1.1.3. and more specifically to Sub Areas 1.1.3.i. *Nanotechnologies and Nanosciences.* Spin Electronics and Nanomagnetism are mentioned in several articles of the *"European White Book on Fundamental Research in Materials Science"*, and in particular research on hybrid magnetic structures, which will be central to our action, is strongly advised.

However, Spin Electronics, by its foreseeable development towards a wealth of nanodevices such as sensors (for automotive applications, biochips, virtual reality games.), ultrahigh capacity mass storage, non volatile magnetic memories for mobile consumer electronics, non-volatile low energy magnetic logic and commutation, etc., pertains also to Thematic Area 1.1.2. and, in particular, to Sub Area 1.1.2.iii. *Components and Microsystems.*

Furthermore, this area requires an increasingly broad range of expertise that encompasses the very latest preparative techniques, patterning of features at the nanoscale, and a wide range of sophisticated analytical and measurement tools all underpinned by sound development of the theory to aid understanding of the fundamentals. Although this exists at the highest level within Europe, it is fragmented into different countries and labs, and current RTN and RTD projects cover too narrow aspects.

If this arrangement may work for the next generation of devices (2005), the future will require much more multidisciplinarity. Already, at the level of training, many RTNs feel the need to hold joint summer schools in order to cover a broader scope.

Hence to maintain the state of the art capability and knowledge base built–up in Europe over the past several years and to be able to expand this and build a platform for a competitive industrial base as the world rises to the opportunity offered by Spin Electronics requires concerted action at the European level as contributed to by the Spin-Net network.

EXCELLENCE

Critical Mass Required and Multidisciplinary Skills

Five main areas of expertise need to be combined to advance the fundamental knowledge of Spin Electronic and Nanomagnetism on the long term:

A. *Materials:* Research and structural characterization on all materials related issues. Major tasks: metallic multilayers, magnetic semiconductors, half metallic ferromagnets with 100% conduction electrons spin polarisation, hybrid systems deposition, *barrier and interface preparation (oxides, nitrides, Schottky barriers) for spin tunnelling and injection.*

B. *Nanofabrication:* Major tasks: development of new high resolution patterning techniques, in an effort to increase the network nanofabrication potential, self organized growth of nanostructures, development of innovative patterning approach that will allow integration into devices.

C. *Functional Properties:* State of the art characterisation and experiments: light emission devices *for* characterisation *of spin injection and spin-optoelectronics*, ultrafast spin and charge dynamics up to fs laser assisted dynamics, *testing spin coherence and dephasing compared to charge relaxation*, high resolution magnetic imaging (special attention will be devoted to magnetic imaging with nm or even atomic resolution), all synchrotron radiation related magnetic characterization and microscopies.

D. *Theory:* A key asset of the network in supporting research efforts of experimental teams. Major tasks: Realistic ab-initio calculations for spin-dependent tunnelling transport, understanding the spin injection process, deciphering exchange mechanisms, Curie temperatures and transport properties of magnetic semiconductors, calculations of magnetic anisotropy and magnetic interactions of nano-structures, current induced magnetisation switching, Ab-initio prediction of new half-metallic systems.

E. *Evaluation of New Concepts of Devices:* Proposing, evaluating and testing all new devices that will emerge. Two major tasks: 1) nanodevice testing: *transfer functions, noise studies, spin accumulation, spin manipulation by gate voltages (Rashba effect, g-factor control), 2)* superposition of coherent spin packets for testing quantum information processing.

The Excellence in these areas of expertise definitely exists throughout Europe, including many pioneers of the latest developments. The critical mass of researchers is also there, for instance national meetings held annually in France or Germany on the subject are always attended by 100 to 200 people. Furthermore, the network SPIN-NET can be readily put together because most of it is already partially organized into European RTN or RTD projects, which will provide a strong basis to build a wider integration.

Proposed Consortium (Organisation, Background and Role)

A detailed presentation of the consortium can be found elsewhere on this web site.

The core of the network is build around partners at the forefront or research, m*ost of whom holding or having held coordination roles at European or National level.* Together they cover most of the scientific expertise needed in the network. The core has 20 nodes distributed over 11 countries, and represents about 200 permanent researchers. A group of associated partners will participate systematically to all actions of the network.

The total will represent about 350 permanent researchers distributed on 35 nodes from 14 countries. Because many contacts already exist within EU or national initiatives, it is estimated that more than 1000 researchers (including PhD students and post-docs) acrossl Europe could be in impacted by our initiative, in particular through the calls for proposals that to be published annually. We believe this represents more than 50% of the work force in this area.

Research Activities

The Network will drive its research activities through the introduction of Task Forces, as the experience of industry tells. Multidisciplinary Task Forces, based on all areas of expertise, will be created to tackle the short term objectives given priority by the Network, and will last as long as necessary to reach them. Based on today's knowledge examples of such objectives could be:

- Operational temperature magnetic semiconductor, compatible with standard 'around-Si' process technology
- High spin polarization materials and contacts (spin-injection / high TMR).
- Spin-manipulation in low-dimensional all-electrical and electro-optical structures
- Magnetic nanostructures by self organisation or non-lithographic methods possibly combined w. lithography
- Ballistic Magnetoresistance technology (magnetic nanocontact technology)
- Thermal stability control in nanostructures
- High speed dynamics (coherence, dephasing, fs laser assisted or probed processes)
- Switching by spin transfer from spin polarized currents
- Magnetic imaging on the sub -10nm scale, with good time resolution wherever possible

Such research priority themes continuously determined in the NoE lead to the realisation of the *long term objectives*, which we consider as key issues for the Future (cf "Expected results" above):

- Obtain magnetic nanostructures with controllable/ switchable / reconfigurable interactions performing complex functions fast and at low power.
- On-chip integrated self assembled magnetic nanostructures and other non-lithographic methods.
- Ballistic Magnetoresistance in perfectly controlled nanocontacts.

- Thermally assisted ultrafast switching on-chip
- Mastering Quantum interference effects of (coherent quantum) magnetic states

INTEGRATION AND STRUCTURING EFFECT

Contribution to the European Research Area

In order to sustain and strengthen the presence of European companies in this field, and at the same time enhance the creation of SME's in the area of spin-based system applications, a strong NOE with a highly interactive mission is essential. It will stimulate the companies to make their investment and job creation in Europe, embedded in the supporting basic research environment our NoE will offer. (Cf. the recent ST/Philips/ Motorola agreement to settle in Grenoble). This is backed by two characteristics of crucial importance:

- Concerted actions on basic research in nanotechnology can boost its implementation in industry.
- Multidisciplinarity is key to the Spintronic nanodevices, extending well beyond the expertise of any single laboratory or even country.

Whilst Europe will benefit in general from such a development, it will help to solve several related problems:

- overall technology standards for research will be enhanced by developing and sharing high quality equipment and by training young researchers at a European level.
- tighter connections to industry will improve the mutual awareness leading to a higher level of financial support from industry, possibly evolving in industrial support for our NoE in future.
- keeping young scientists in Europe and attracting some of the best foreign ones by the quality

opportunities that our integrated effort, connected to industry, will offer.

Integration Activities

SPIN-NET will be established as a *"virtual laboratory"* at the scale of the whole network through:

- The constitution of a Governing Council and of a Scientific Committee,
- Network-wide priority topics and updating these according to a joint road-map
- Promotion and support of collaborative actions
- identification and involvement of new partners
- Accessibility to the very diverse expertise of the network, e.g., a common web site (www.spin-net.org), partly public access with features as: chat room for scientific problems, new partners identification, connect young researchers, etc.
- common services, e.g.: sharing resources and dedicated equipment, fund easy access to nanofabrication centres, etc.
- enhance human resources in our community and increase flexibility and (multicultural) integration aspects
- organise publications (newsletter), PR, IPR and liaisons to public authorities, media, industry.

Special actions will be taken to help groups from the NAS to upgrade their equipment and training to reach state of the art level (partners from Poland and Czech Republic already are in our network core, and partners from Hungary and Coatia are associated).

Dissemination Activities

The groups present in this network have an excellent track record regarding scientific publications. Some core members are in the editorial boards of leading journals, enabling them to flag to their colleagues major changes in the field as identified in the NoE. *The NoE members are all strongly involved in the organisation of the National and International conferences in the field. One of the actions that might be taken is a common European action to avoid overlap in topics at various meetings.*

Road map actions will provide input to EU and National Policy Committees.

The academic institutes in the NoE will compare the content of Master course lectures and exchange lecturers on that level within the NoE. Further, the high standard of organizing summer schools etc enhances the number and level of Educational Courses

A Vice-coordinator will be appointed in charge of the communication directly to users of the results through Policy-making workshops, web communities of excellence, email groups, training courses, eLearning initiatives, Guidelines, Best Practices. Where appropriate Regional Development Organisations e.g. Innovating Regions of Europe, BICs, IRCs, etc. will be approached within the above mentioned policy making workshops.

The Role of Industry in Spin-Net

Spin Electronics has such a high potential for applications that it is necessary to keep a very close contact with representatives of the European industry. Organization will proceed as follows:

- some companies will participate in the network and the research as full partners
- some will have associated partner status (mainly companies that have less of a tradition in long term basic research) - with only travel support to participate to workshops, training sessions and meetings

representatives of high technology SMEs and larger size companies, will constitute a large part of the Committee contributing towards the definition of the research orientations of the network

Using modern epitaxy methods, it is now possible to control the growth of semiconductor crystals with atomic-layer precision. In particular it is possible to incorporate clean "2-dimensional" planes and arrays of magnetic material into otherwise nonmagnetic semiconductor lattices. In this way the physics of (possibly interacting!) low-dimensional magnetic layers can be investigated.

All of the structures are made by our long-time collaborator Professor Nitin Samarth in the Department of Physics at Pennsylvania State University. Much of our work with Professor Samarth involves the study of the dynamical interplay between photoexcited electronic carriers (electron, holes, and excitons) and local magnetic moments (Mn, for instance) that are embedded in a semiconductor quantum structure.

The carriers and local magnetic moments couple to one another via their respective magnetic spins, and the resultant spin-scattering occurs on femtosecond and picosecond timescales which require ultrafast laser techniques to resolve. Two such techniques are time-resolved Faraday rotation, which is sensitive to the net magnetization (spin) of the system under study, and time-resolved upconversion of photoluminescence, which reveals the picosecond evolution of photoexcited spin-polarized carriers.

A particularly attractive host in which to study these spin dynamics is a newly-developed class of magnetic semiconductor called digital magnetic heterostructures. In these, a fixed amount of magnetic material can be 'digitally' incorporated in different configurations (within a nonmagnetic quantum well, say) to study the effects of interfaces and reduced magnetic clustering on the underlying spin-scattering processes. Recent experiments have demonstrated the ability to observe terahertz spin precession and ultrafast transfer of angular momenta in these digital structures. These phenomena now enable all-optical spin resonance of single monolayers of magnetic semiconductor.

Digital magnetic heterostructures incorporate discrete monolayer and sub-monolayer planes of purely magnetic semiconductor (Manganese-Selenide in this case, shown below in green) into otherwise nonmagnetic ZnSe-ZnCdSe-ZnSe quantum wells. In this way one can directly control, using a fixed number of magnetic spins, such factors as the exciton wavefunction within the well (shown in red), the local in-plane spin-density (affecting cluster formation), and the overlap of the wavefunction with the magnetic planes.

Three such digital structures are shown above (the schematics are of bandgap energy versus growth direction). Each incorporates only 3 monolayers of magnetic material, but in systematically varied distributions. The effect of this increased distribution is to radically increase the Zeeman splitting (see below) and the measured spin-scattering, and to decrease the statistical occurrence of antiferromagnetically-locked Mn clusters. In the presence of an applied magnetic field, the planes of Mn become magnetized.

The magnetic barriers are thus lowered for spin-down electrons and holes (excitons), and are increased for spin-up electrons and holes. Alternatively, spin-up excitons "see" a different barrier height than their spin-down counterparts. The net effect is to energy-split the exciton spin-states, shown schematically below. In these magnetic heterostructures, this Zeeman splitting can be made hundreds of times larger than in nonmagnetic quantum structures.

In the absence of an applied magnetic field, spin-up and spin-down states are degenerate, and their spin relaxation may be seen as mirror images of each other (black curves), with a relaxation time in this sample of ~20 picoseconds. Upon the application of a magnetic field, the results could not be more different: the energy degeneracy is lifted, and the coherently-excited spin polarizations exhibit quantum beating between the Zeeman-split levels at terahertz frequencies, with a frequency proportional to the electron g-factor and a rapid dephasing time.

Although the electronic dynamics disappear after ~20 picoseconds, more careful examination of the data shows that the system has not returned to equilibrium. A second signal is seen with a different period, corresponding to a g-factor of 2, and is a signature of the manganese ions. Angular momentum is coherently transferred from the electronic system to the magnetic spins: unlike traditional semiconductors, the electronic carriers have left a "footprint" in the magnetic sublattice which subsequently persists long after the carriers have disappeared from the structure. Although it has returned to electronic equilibrium, the quantum well is still undergoing dynamical spin processes.

This magnetic spin precession persists over far longer timescales - hundreds of picoseconds - eventually relaxing due to phonons in the system. These types of studies allow us to disentangle electronic and magnetic spin phenomena in quantum structures, enabling us to perform spin resonance experiments in systems with near atomic thicknesses.

ELECTRONIC STATES IN MAGNETIC NANOSTRUCTURES

Why Nanostructures?

The popularity of the affix *nano* begs the question why the physics and technology of nanostructures should be special. Are there phenomena occurring at the nanometer scale that are not encountered in the more familiar micro-world? A rather global answer concentrates on two related phenomena: 1) Nanometer dimensions are comparable to the electron wavelength in a solid. 2) When electrons are confined to dimensions comparable to their wavelength, the continuum of bulk energy bands becomes quantized into discrete quantum well states.

The energy spacings of the states increase as the structures become smaller, eventually becoming greater than the thermal energy kT. If that happens for states at the Fermi level E_F, only a single quantum level is accessible thermally. Consequently, the dimensionality of electrons is reduced in a nanostructure because they cannot propagate along the directions of the confinement. These effects can be illustrated by analyzing the electronic states in thin metal overlayers by photoemission and its time-reversed counterpart, inverse photoemission.

The two techniques are particularly useful for probing electronic states, since they make it possible to determine the complete set of quantum numbers that characterize an electron in a crystal, i.e., energy E, momentum p (or wave vector k = p/), angular symmetry, and spin. Photoemission measures occupied states, inverse photoemission unoccupied states. We focus on electronic states near the Fermi level which are essential in transport properties, such as magnetoresistance, and drive magnetic transitions. For a recent review of magnetic nanostructures, see.

The density of unoccupied states at E_F with parallel momentum k = 0 is displayed in Figure 1 for Cu films grown epitaxially on an fcc Fe film which, in turn, was grown epitaxially on a Cu(100) single crystal. As the thickness of the Cu overlayer is increased, the density of states oscillates with a period of about six atomic layers (about 1 nm), which brings us into the nanometer regime. The amplitude of the oscillations is significant—there is no zero offset in Figure 1. In fact, the amplitude appears to be limited by the smoothness of the interfaces that is achievable by current growth techniques. The smoother the films, the larger the amplitude. This explains why such oscillations have been difficult to observe in thin films so far.

The density-of-states oscillations can be understood by a simple interferometer model (Figure 1, top). Reflection of electrons at the interfaces builds up standing waves whenever the round-trip phase is a multiple of 2. The resulting interference fringes appear with a period of / 2, where is the wavelength of the electrons. Thus, we are measuring the wavelength of electrons directly in real space, using the world's smallest interferometers. In the case of Cu(100) at $E = E_F$ and k = 0, we obtain a wavelength of 2 nm. It might be puzzling why this value is so much larger than the Fermi wavelength, which is comparable to a lattice constant. What is measured here is the wavelength of an envelope function that modulates the normal Bloch function in order to satisfy the boundary conditions at the interfaces.

To see the connection between density of states and quantum well states, we have to look at the whole spectrum of electronic states (Figure 2). For a bulk Cu(100) crystal (top spectrum), a continuous spectrum is observed which corresponds to the *s, p* band with $_1$ symmetry that straddles the Fermi level and has an upper cutoff at 1.8

eV above E_F, which corresponds to the p_z-like X_4' point of the Cu band structure. In the thin-film spectra, which were obtained from overlayers differing in Cu thickness by two layers from one curve to the next, the continuous spectrum breaks up into discrete quantum well states.

These states change their energy with decreasing Cu thickness and cross the Fermi level at regular thickness intervals. These Fermi-level crossings give rise to the density-of-states maxima in Figure 1. The fact that quantum well states change their energy with thickness is easy to understand by electron counting. In a finite slab of N atomic layers, one has N states at each k point. Therefore, the energy interval between adjacent states shrinks as $1/N$ with increasing film thickness and approaches zero for the bulk. For the very thin films in Figure 2, the spacings are still large compared to $kT = 0.026$ eV (at room temperature): e.g., five unoccupied states within 2 eV for a 20-monolayer film, giving an average spacing of 0.4 eV. We would have to make the films 15 times thicker (about 50 nm) to reach the point at which the energy spacing becomes comparable to kT. Again, these dimensions place the region of interest in the nanometer regime.

Oscillatory Magnetic Coupling and GMR

The quantum well states seen in Figures 1 and 2 are closely correlated with magnetic properties such as oscillatory magnetic coupling and giant magnetoresistance (GMR). Figure 3 compares the density of states, spin-polarization, and magnetic coupling for the fcc Cu/Co(100) system. All three quantities are found to oscillate with the same period of about six atomic layers (1 nm). Giant magnetoresistance in trilayers exhibits the same period, since it is tied to antiparallel coupling. Such an agreement is found wherever comparable data exist, e.g., for Ag/bcc

Fe(100), Au/bcc Fe(100), Cu/fcc Co(100), and Cu/fcc Fe(100). For the short-period oscillations in Cu/fcc Co(100), observation of this effect requires highly perfect interfaces.

Recently, density-of-states oscillations have also been seen for transition-metal spacers: i.e., Cr/bcc Fe(100). After seeing such a correlation in periodicities, one would like to make the physical connection between quantum well states and oscillatory magnetic coupling. This can be done by calculating the total energy of all occupied quantum well states and observing it jump periodically with increasing thickness whenever a new state crosses the Fermi level. Here we restrict ourselves to a qualitative discussion, using a suggestion[1] that explains why antiparallel coupling correlates with a density-of-states maximum at the Fermi level E_F in Figure 3. A high density of states at E_F is energetically unfavorable.

Thus, the magnetic arrangement will undergo a restructuring to minimize the density of states at E_F. In a trilayer, confined quantum well states can occur only for parallel orientation of the magnetic layers, as shown in Figure 4(a). For an antiparallel orientation, the wave functions extend into one of the magnetic layers. Therefore, a transition to antiparallel coupling circumvents the high density of states created by a quantum well state.

Following this argument, it might be puzzling that quantum well states are observed in Figures 2 and 3 for spacer thicknesses characteristic of antiparallel coupling. The density-of-states data are not for trilayers, however, but for Cu/Co(100) bilayers, where one of the ferromagnetic layers is replaced by vacuum, which always confines the wave function. It is difficult to perform such surface-sensitive experiments for trilayers, where the ferromagnetic overlayer tends to swamp the signal from the spacer. First results on trilayers demonstrate that it is possible to penetrate the overlayer under favorable conditions.

Quantum well states must be spin-polarized for transmitting magnetic information, even in a noble-metal spacer such as Cu. This rather counterintuitive picture of a "magnetized" noble metal has been confirmed by spin-polarized photoemission (Figure 3, middle) and by magnetic circular dichroism. There is a rather simple and general explanation for the spin-polarization of quantum well states in magnetic multilayers. Since they are formed by reflection of electrons at the interfaces, any spin-dependence in the reflectivity affects their confinement. In particular, the reflectivity at a ferromagnet differs between majority and minority spins, because of the magnetic exchange splitting of the bands.

A particularly clear-cut case is shown in Figure 4(b) for the interface between fcc Ag(100) and bcc Fe(100) at k = 0. Minority spin states at E_F in the Ag layer have no counterpart in Fe. Therefore, they are totally Bragg-reflected. Majority states, on the other hand, exist on both sides of the interface and are able to propagate. This leads to quantum well states with minority spin character, which have been observed by spin-polarized photoemission. They are superimposed on a continuum of majority spin states. In general, there is not such a clear distinction between the reflectivity of majority and minority spins.

However, one can argue that there will always be a spin-dependent potential step at the interface that makes the reflectivity spin-dependent. For spacer layers to the right of the ferromagnets in the periodic table, the majority bands are more closely aligned than the minority bands because of a better match in band filling. This explains the observation of quantum well states with minority spin in noble metals. For spacers to the left of the ferromagnets, e.g., Cr, the situation is reversed, and one expects quantum well states with majority spin character. There are indications of quantum well states in Cr, but their spin-polarization has not yet been measured. This argument holds independently of the specific band topology, but it does not guarantee full confinement, which would require 100% reflectivity at the interface.

Spin-dependent reflectivity of electrons at interfaces is not only important for quantum well states and magnetic coupling, but might have an effect on giant magnetoresistance as well. Models of the magnetoresistance for "conductivity in-plane" (CIP) and "conductivity perpendicular to the plane" (CPP) provide spin asymmetries for interface and bulk scattering in magnetic multilayers.

Typically, interface spin asymmetries are four times larger than in the bulk, which leads to the fact that the CPP geometry exhibits 3-10 times higher magnetoresistance than the CIP configuration. It should be kept in mind, though, that these asymmetries could be due to diffuse scattering instead of the specular reflection discussed here, and that many other parameters enter into the equations for magnetoresistance.

Concentrating on specular interface scattering in the CPP geometry, one can construct a simple polarizer-analyzer model for giant magnetoresistance: Each interface with a ferromagnet acts as a polarization filter for electrons. If two magnetic layers are aligned antiparallel, one has the analog of crossed optical polarization filters, i.e., little transmission. If they are parallel, they both transmit electrons. The smaller magnetoresistance of the CIP geometry can be understood as a shunting effect whereby electrons are able to avoid spin-dependent scattering at the interfaces by traveling inside the layers.

In order to achieve the high magnetoresistance of the CPP geometry but avoid its extremely low absolute resistance (μ and less), several attempts have been made to produce one-dimensional wire or thin-film stripe structures, in which the current runs perpendicular to the interfaces but is confined to the wire or the thin film of stripes. Lithographic patterning has been used to produce sawtooth-shaped Si substrates for the deposition of magnetic stripes. Magnetic wires perpendicular to the surface have been obtained by filling pores in a polymer film with electroplated metals. Designing such structures opens up a fascinating world of low-dimensional growth phenomena. In the following section, we discuss a new approach for producing stripes with nanometer width at stepped surfaces.

Since electron reflectivity at interfaces plays an important role in the magnetic phenomena discussed here, it is natural to try modifying the phenomena by coating the interfaces with a different material. Using one or two monolayers of a ferromagnet with large magnetic band splitting, such as Co, it has been possible to enhance the magnetoresistance of permalloy/Cu/permalloy structures by more than a factor of 2. A variety of special electronic states have been found at such "doped" interfaces by inverse photoemission, e.g. in Cu/Co/Ni(100).

A Co3*d*-like interface state builds up at the same fast rate as the enhanced magnetoresistance, suggesting a connection with the magnetic band splitting in Co, which is three to four times as large as in Ni. (The spin splitting is about 2 eV in bcc Fe, 1 eV in fcc Fe, 1 eV in fcc and hcp Co, and 0.3 eV in fcc Ni.) Quantum well states in the subsequent Cu layer are enhanced near the interface, indicating stronger confinement by higher, spin-dependent reflectivity. Going to thicker ferromagnetic overlayers, one finds quantum well states not only in the noble-metal spacer, but also in the ferromagnet. Returning to the optical interferometer analogy, one now has reflective or antireflective coatings on the electron mirrors, and the reflectivity oscillates with the thickness of the ferromagnetic overlayer. This affects the strength of the magnetic coupling, but not its periodicity, which is determined by the noble-metal spacer layer.

Stripes and Dots at Stepped Surfaces

For growing one-dimensional nanostructures we consider growth at stepped surfaces, where the steps provide a template for attaching magnetic wires or stripes. In the two-dimensional world of the first monolayer growing on a stepped surface, one can observe growth modes analogous to layer-by-layer and Stranski-Krastanov growth in three dimensions. As shown in Figure 5, Cu grows on a stepped Mo(110) surface in parallel stripes that correspond to a row-by-row growth mode. On W(110), however, only the first row of Cu atoms decorates the step edge, and additional Cu grows in monolayer-height islands that are attached to the step edges, analogous to Stranski-Krastanov growth.

An additional phenomenon at stepped surfaces is the presence of an energy barrier for crossing steps. This is obvious for an uphill crossing, since the atom incorporated at the step edge would be less coordinated on the terrace. Even for a downhill crossing, the atom must temporarily give up neighbors when crossing the step edge. An atom exchange mechanism can eliminate this barrier on certain metals during homoepitaxy. In a heteroepitaxial system, such as Cu/Mo(110), the Cu atom crossing a Mo/Cu boundary trades Mo neighbors for Cu neighbors, which provide less binding energy. Such a barrier causes the inhomogeneous width of the Cu stripes on Mo(110) in Figure 5(b). The wider the terrace on which a Cu stripe resides, the wider the stripe. Apparently, the Cu atoms migrate toward the uphill edge of an individual terrace. It requires higher annealing temperatures to let Cu atoms cross step edges and equilibrate their stripe width, independent of the terrace width. Stripes with a uniform width of 3 nm have been obtained for Cu on stepped Mo(110) after annealing to 600°C. The existence of uniform stripes in spite of nonuniform terraces indicates that the binding energy of Cu increases monotonically toward the step edge.

To make step decoration a widely applicable technique, it has been proposed that a two-step process be used (Figure 6): In the first step, a template of inert stripes is grown by step flow. In the second step, the desired material is deposited selectively on the remaining stripes of the reactive substrate using chemical vapor deposition (CVD). For example, one could consider a stepped Si(111) surface as substrate, grow passivating stripes of lattice-matched CaF_2 along the step edges, and deposit metals by selective CVD on the remaining Si. Compounds for substrate-selective CVD have been developed for contacting Si devices through via holes in SiO_2 without shorting out the SiO_2 insulation. Selective deposition of W or Ta via WF_6 or TaF_5 might be useful to create a reaction barrier on the Si stripes against the formation of silicides with Fe, Co, or Ni. For the deposition of noble-metal spacers, such as Au, there exist organometallic precursors. The spacers could also be simply evaporated after selective CVD of ferromagnetic wires.

If it becomes feasible to grow magnetic nanowires by step decoration, one could go one step further and produce an array of regular dots by patterning the wires lithographically (Figure 7). Using a UV laser, a standing-wave pattern with a periodicity of 200 nm can be produced on a photoresist that would allow elongated dots with typical dimensions of 20×200 nm^2. Such a regular array of highly anisotropic magnetic particles might make a good medium for the ultimate goal in magnetic recording, single-particle-per-bit storage.

Current media with randomly segregated particles in a CoPtCr alloy typically use 10^3 particles per bit to reduce the readout noise introduced by irregular particles and domains. Working with a small subarray of magnetic dots per bit (say 1×100) already reduces this large number significantly without deviating too far from the traditional longitudinal recording geometry. When the number of dots per bit becomes too small, however, one is likely to run into tracking problems. Other geometries are conceivable that preserve a reasonably wide track, e.g., the two-particle-per-bit transverse recording scheme shown in Figure 7. The stray field between antiparallel-magnetized pairs of elongated dots would become the readout signal. The two bit levels would correspond to magnetizations pointing toward each other or away from each other.

Element-sensitive STM via Resonant Tunneling

For determining how stripes grow at steps, it is essential to have a high-resolution microscope that not only resolves the topography but is also able to distinguish between the substrate and the new material incorporated at the step edges. For perfect step-flow growth of lattice-matched materials, there is no topographic contrast. To achieve chemical contrast, we have developed an element-specific version of scanning tunneling microscopy (STM). It is based on resonant tunneling into empty surface states that are specific to each metal.

Thus, the electronic structure of nanostructures again becomes important. Figure 8 shows two STM images of the same three Cu stripes (about 3 nm wide) on a stepped Mo(110) surface. At specific bias voltages, it is possible to make the Cu appear either brighter than Mo (at +5 V) or darker (at +0.06 V). Such chemical contrast results from resonant tunneling of electrons from the tip into unoccupied surface states of the Cu or Mo, respectively. With most other bias voltages there is little contrast between the Cu and Mo (not shown). This reflects the topography, since Cu has practically the same size as Mo and grows in registry with the substrate lattice.

The states available for resonant tunneling can be found with inverse photoemission (Figure 9, shaded states). There are two types of element-specific surface states for Cu on W(110) and Mo(110). One is a p_z-like surface state near the Fermi level, the other an image state near the vacuum level. Mo appears bright at a bias of +0.06 V because of tunneling into its p_z-like surface state; Cu appears bright in Figure 8 at +5 V bias because of tunneling into its image state. While the upward shift of the p_z-like surface state from W to Cu is nontrivial, the downward shift of the image state is easily understood. It tracks the work function reduction from W to Cu, since image states are bound to the vacuum level E_v and the energy zero is the Fermi level E_F in inverse photoemission and STM ($= E_v - E_F$). The work functions of metal surfaces are well known and have been tabulated, allowing straightforward identification of different metals at a surface. Such tunneling resonances into image states are being modeled with realistic tip geometries.

To quantify resonant tunneling, we display the enhancement of the Cu stripes in the constant-current mode in Figure 10. The apparent height of the Mo-Cu step is normalized to the actual, Mo-Mo step height of 0.2 nm. At bias voltages corresponding to the $n = 1, 2,$

3 image-state resonances, we find sharp resonances with enhancements of up to 100%. If we note that the tunneling current increases typically by two orders of magnitude over such a distance in STM, we expect a huge effect on current images at constant height.

This large enhancement is due to the narrow intrinsic width of the image states, which can be as low as 0.02 eV in Cu. The p_z-like surface state is broader and generally does not produce such a large effect, but it allows higher spatial resolution, since the tip comes closer to the surface at low bias voltages. For example, the narrow parts of the Cu stripes in Figure 8 are resolved in the high-resolution image at +0.06 V but are lost at +5-V bias. We have found it useful to first identify the low-work-function material by its $n = 1$ image state and then, by tuning to a low-bias surface state, proceed to a higher resolution.

Outlook

This brief overview has illustrated the key role played by electronic states in magnetic nanostructures. The states are connected with interesting effects observed in these structures, such as oscillatory magnetic coupling, giant magnetoresistance, and interface doping. They are also essential for imaging the growth of nanostructures by resonant tunneling, e.g., imaging the growth of metallic stripes at stepped surfaces. The experiments with such self-assembled, one-dimensional structures are typical for a new wave of "engineered" materials.

Magnetic and transport properties can be altered dramatically by structuring materials on a scale comparable to the electron wavelength, as demonstrated by noble-metal layers that become magnetized when sandwiched between ferromagnets. There is ample flexibility for designing magnetic structures that are customized for applications such as magnetic storage media or sensors.

NANOBIOMAGNETICS

"Nanobiomagnetism" is the intersection of nanomagnetism and medicine that focuses on biological systems and/or processes. Magnetism is an inherent facet of life, from iron in blood to the ability of magnetotactic bacteria, birds, honeybees and other creatures to navigate by the Earth's magnetic field.

Iron plays a critical part in many aspects of human neurophysiology. Naturally occurring iron in the body usually is stored within ferritin, which are 12-nm hollow spherical shells that each can hold up to 2,500 iron atoms in the form of mineralized ferrihydrite. Anomalous amounts of iron—possibly in nanoscale form—are associated with many neurodegenerative disorders such as Alzheimer's, Parkinson's, and Huntington's diseases. The ability of magnets to act on objects at a distance makes them valuable medical tools. A 1624 report described the extraction of an iron splinter from an eye using a magnet.

Safety pins, bullets and grenade splinters were removed using magnets. Grazing cows are fed magnets to prevent sharp metallic objects they eat from damaging the intestines. The invention of stronger, smaller permanent magnets made possible more delicate applications, such as temporarily fixing prosthesis in dentistry, guiding catheters through the body, and navigating within the brain. Nanoscale materials have a special relevance to biomedical applications due to their size compatibility with cells (10–100 ìm), viruses (20–450 nm), proteins (5–50 nm) and genes (2 nm wide by 10–100 nm long). Nanoparticles are small enough to move inside the body without disrupting normal functions, and can access spaces inaccessible by other means. Cells react to the topography of their environment on size scales as small as 5 nm - up to 1000 times smaller than their own size. Changes in response to topography literally can induce growth or death. Nanostructured materials allow study of these critical processes on a single-cell level.

Materials

Magnetic biomaterials have different constraints than materials used for other applications. *In vivo* (in the body) applications require strict biocompatibility. *In vitro* (outside of the body) applications have less strict requirements, but techniques involving living cells still must consider the effect of the materials on the sample under study. In addition to biocompatibility, materials must be capable of being functionalized with one or more molecules, must retain their magnetic properties for a reasonable period of time in aqueous media with varying pH, must not be cleared too quickly from the bloodstream, and must form stable, non-aggregating dispersions.

Biotoxicity

Cells can be killed by external agents or can be induced to "commit suicide"via a pattern of events called programmed cell death or *apoptosis*. Injuries from external agents include mechanical damage and exposure to toxic chemicals (e.g. chemotherapy). Substances toxic to cells are called cytotoxic, and different types of cells can have different responses to the same material. The vast majority of research in this area involves iron-oxide particles, as they are highly biocompatible, cheap, can be made in a variety of ways and sizes, and can be made as superparamagnets or ferrimagnets. Iron oxides are metabolized into elemental iron and oxygen by hydrolytic enzymes in the body.

The iron joins the normal body stores and the body compensates by taking up less iron from the stomach. Intravenous injections of up to 250 mg iron/kg body weight does not produce chronic or acute hepatotoxicity in rats, while 1–3 mg/kg has been used clinicallyin humans.

The disadvantage of iron oxides, however, is their low magnetic susceptibilities. Iron nanoparticles offer an order of magnitude greater susceptibility than iron oxides at room temperature, but are easily oxidized and are not as biocompatible as their oxides. Fe-C composite particles made by mechanical milling, chemical reduction, or plasmochemical recon-densation have been used in clinical trials for hyperthermia and drug delivery.

These particles are biocompatible and the carbon in the particles may assist in reversing drug-induced toxicities via physical adsorption. The need for monodisperse particles for magnetic recording has stimulated a many chemical methods that produce a broad variety of element and alloy nanoparticles, including Co, Ni, Cu, FePt, CoPt, Fe, $CoFe_2O_4$, $MnFe_2O_4$, $SmCo_5$ and even core-shell particles such as FePt/Fe3O4. Physical deposition techniques such as inert gas condensation and laser ablation also are used.

Coatings

Coatings can improve oxidation resistance, colloidal stability, the ability to functionalize, phagocyte resistance, mechanical stability, and biocompatibility. Unfortunately, coating a magnetic nanoparticle with a biocom patible material does not necessarily render the nanoparticle biocompatible. The chemistry for functionalizing gold is well established and gold coating offers corrosion resistance; however, biotoxicity issues may need to be resolved. Polysaccharide coatings such as dextran, starch, and chitosan are biocompatible and offer a range of functionalization options; however, they can be structurally weak and can be dissolved by highly acidic environments.

Silicon-based coatings are used to protect particles from lysomal enzymatic digestion, and improve mechanical pro perties and chemical stability. Silica-coating can improve chemical stability, but a porous coating may allow the contents inside to be dissolved or oxidized. Many polymers are biocompatible and may be used as coatings for metallic or ceramic particles, or can serve as hosts by either capturing nanoparticles inside a larger polymer particle or attaching nanoparticles to their surfaces. Polyethylene Glycol (PEG) and related polymers covalently bond to surfaces or are adsorbed on magnetic nanoparticles and can prolong the circulation time in the bloodstream.

Clearance Time

Particles introduced into the bloodstream are covered rapidly by components of the circulation, such as plasma proteins, in a process called *opsonization.* Opsonization makes the particles recognizable to the body's major defense system, the reticuloendothelial system (RES). The RES comprises a diffuse system of phagocytic cells (which engulf inert material) that are primarily associated with the connective tissues in the liver, spleen, and lymph nodes. Macrophage (Kupffer) cells in the liver and macrophages of the spleen and circulation are important in removing particles identified by opsonization. A significant fraction of nanoparticles can be cleared from the circulation system in as little as 15 minutes.

The clearance rate is dependent on size, charge, surface hydrophobicity and the number and nature of functional groups on the surface. These variables are interdependent, making understanding the role of each one independently challenging. Some of these variables also may affect the magnetic properties. For example, smaller particles more easily evade the RES; however, the smaller size usually results in a smaller moment.

Anionic particles with negative surface charge have a high affinity for the cell membrane and are typically taken up by the endocytic process. Cationic magnetite particles show significantly lower cell-survival rates but their toxicity depends highly on the magnitude of the surface charge. More highly cationic particles tend to be more toxic. Hydrophilic surfaces such as dextran, polyethylene glycol, polyethylene oxide, poloxamers, polysorbates and polyoxamines provide a dynamic "cloud" of hydrophilic and neutral chains at the particle surface that repel plasma proteins and prevent rapid removal of particles from circulation. Related to clearance time is the manner in which the nanoparticles attach to cells. Nanoparticles can be internalized or remain adhered to the surface. The mechanism is determined by the surface charge, adhesion properties, and chemical functionality of the cells with respect to the nanoparticle.

Other Applications

Magnetic nanoparticles can be used simultaneously for more than one of the applications discussed; for example, MRI can be used to confirm magnetic nanoparticle distribution prior to using the same particles to administer hyper-thermia. A combination of fluorescence and magnetic-nanoparticle-enhanced MRI was used for preoperative magnetic resonance imaging and as an intraoperative optical probe during surgery, allowing clearer delineation of brain tumors. These "multimodal nanoparticles" may allow radiologists and neurosurgeons to see the same probe in the same cells and thus improve identification of tumor margins. A nanoparticle combining near-infrared fluorescent dye with magnetism allows the particles to be located by MRI while the fluorescent dye provides simultaneous information about the molecular environment about the nanoparticle.

Although most applications try to avoid agglomeration, the intentional formation of a blockage of magnetic nanoparticles to block blood supply to a tumor has been investigated. Aggregation and selective uptake has been used to destroy cells via the application of pulsed magnetic fields and subsequent rupture of cells. Retinal detachment is a major cause of vision loss in adults. The usual treatment is the scleral buckle, which is a silicone band sewn to the outside of the eye that compresses the wall of the eye inward to close the holes in the retina.

Sterically stabilized 4–10 nm magnetic particles in a poly(dimethylsiloxane) biocompatible fluid can be held in place with an external magnetized scleral buckle, thus providing a stable internal blockage that encircles the entire eye periphery with a ring of silicone oil. No fluid contacts the central vitreous cavity or the lens, thus decreasing the chances of undesired contact and/or damage. Many neurodegenerative diseases indicate the disruption of normal iron homeostasis in the brain. Recent experimental work indicates that nanoscale magnetic biominerals (primarily magnetite and maghemite) may be associated with senile plaques and fan filaments found in brain tissue affected by these diseases. Understanding the role of iron in neuro degenerative disease could help understand the origin, diagnosis and treatment of these diseases.

CONCLUSION

The field of nanobiomagnetics is exceptionally broad, involving researchers From medicine, pharmaceutical science, chemistry, physics, biology, engineering, and materials science. The literature is similarly decentralized, making it Impossible not to omit some contributions in this article. Recent reviews mentioned in this chapter provide additional information for the reader interested in more detail on a specific topic. Nanomagnetics researchers have much to contribute to this new and exciting field.

Bibliography

Attebery, B.: *Nanoculture: Implications of the New Technoscience*, Bristol, Intellect Books, 2004.

Berube, D.M.: *The Rhetoric of Nanotechnology*, Amsterdam: IOS Press, 2004.

Charles, P., and Frank J. Owens: *Introduction to Nanotechnology*, Hoboken, J. Wiley, 2003.

Drexler, K. Eric: *Engines of Creation: The Coming Era of Nanotechnology*, New York, Anchor Books, 1986.

Glimell, H.: *Dynamics of the Emerging Field of Nanoscience*, Dordrecht, Kluwer, 2001.

Gregory, L.: *Nanotechnology*, New York, Springer, 1999.

Hayles, N.K. : *Nanoculture: Implications of the New Technoscience*, Bristol, Intellect Books, 2004.

Jin Z. Zhang: *Self-assembled Nanostructures*, New York, Kluwer Academic/ Plenum, 2003.

Pressman, J.: *Nano Narrative: A Parable from Electronic Literature*, Bristol, Intellect Books, 2004.

Sarewitz, D.: *Nanotechnology and Societal Transformation*, Dordrecht, Kluwer, 2001.

Schummer, J.: *Interdisciplinary Issues in Nanoscale Research*, Amsterdam: IOS Press, 2004.

Suchman, M.: *Envisioning Life on the Nano-Frontier*, Dordrecht, Kluwer, 2001.

Tenner, E.: *Nanotechnology and Unintended Consequences*, , Dordrecht, Kluwer, 2001.

Weil, V.: *Ethical Issues in Nanotechnology*, Dordrecht, Kluwer, 2001.

Index

□□□